SAT Subject Test

Mathematics

Student Practice Workbook

+ Two Full-Length SAT Math Tests

Math Notion

www.MathNotion.com

SAT Subject Test Mathematics

Published in the United State of America By

The Math Notion

Web: WWW.MathNotion.com

Email: info@Mathnotion.com

ISBN: 978-1-63620-046-0

The Math Notion

Michael Smith has been a math instructor for over a decade now. He launched the Math Notion. Since 2006, we have devoted our time to both teaching and developing exceptional math learning materials. As a test prep company, we have worked with thousands of students. We have used the feedback of our students to develop a unique study program that can be used by students to drastically improve their math scores fast and effectively. We have more than a thousand Math learning books including:

– SAT Math Prep

– ACT Math Prep

– GRE Math Prep

– Accuplacer Math Prep

– Common Core Math Prep

–many Math Education Workbooks, Study Guides, Practice and Exercise Books

As an experienced Math test preparation company, we have helped many students raise their standardized test scores—and attend the colleges of their dreams: We tutor online and in person, we teach students in large groups, and we provide training materials and textbooks through our website and through Amazon.

You can contact us via email at:

info@Mathnotion.com

Get the Targeted Practice You Need to Ace the SAT Math Test!

SAT Subject Test - Mathematics includes easy-to-follow instructions, helpful examples, and plenty of math practice problems to assist students to master each concept, brush up their problem-solving skills, and create confidence.

The SAT math practice book provides numerous opportunities to evaluate basic skills along with abundant remediation and intervention activities. It is a skill that permits you to quickly master intricate information and produce better leads in less time.

Students can boost their test-taking skills by taking the book's two practice SAT Math exams. All test questions answered and explained in detail.

Important Features of the SAT Math Book:

- A **complete review** of SAT math test topics,
- Over 2,500 practice problems covering all topics tested,
- The most important concepts you need to know,
- Clear and concise, easy-to-follow sections,
- Well designed for enhanced learning and interest,
- Hands-on experience with all question types
- **2 full-length practice tests** with detailed answer explanations
- Cost-Effective Pricing

Powerful math exercises to help you avoid traps and pacing yourself to beat the SAT test. Students will gain valuable experience and raise their confidence by taking math practice tests, learning about test structure, and gaining a deeper understanding of what is tested on the SAT Math. If ever there was a book to respond to the pressure to increase students' test scores, this is it.

Contents

Chapter 1 :

Integers and Number Theory

Topics that you'll practice in this chapter:

- ✓ Rounding
- ✓ Whole Number Addition and Subtraction
- ✓ Whole Number Multiplication and Division
- ✓ Rounding and Estimates
- ✓ Adding and Subtracting Integers
- ✓ Multiplying and Dividing Integers
- ✓ Order of Operations
- ✓ Ordering Integers and Numbers
- ✓ Integers and Absolute Value
- ✓ Factoring Numbers
- ✓ Greatest Common Factor (GCF)
- ✓ Least Common Multiple (LCM)

"Wherever there is number, there is beauty." –Proclus

Rounding

Round each number to the nearest ten.

1) 42 = ____
2) 88 = ____
3) 24 = ____
4) 57 = ____
5) 19 = ____
6) 25 = ____
7) 93 = ____
8) 71 = ____
9) 48 = ____
10) 81 = ____
11) 58 = ____
12) 87 = ____

Round each number to the nearest hundred.

13) 198 = ____
14) 387 = ____
15) 816 = ____
16) 101 = ____
17) 321 = ____
18) 433 = ____
19) 579 = ____
20) 825 = ____
21) 580 = ____
22) 868 = ____
23) 480 = ____
24) 287 = ____

Round each number to the nearest thousand.

25) 1,382 = ____
26) 3,420 = ____
27) 4,254 = ____
28) 6,861 = ____
29) 9,099 = ____
30) 22,980 = ____
31) 45,188 = ____
32) 16,808 = ____
33) 52,866 = ____
34) 85,190 = ____
35) 70,990 = ____
36) 26,869 = ____

Rounding and Estimates

Estimate the sum by rounding each number to the nearest ten.

1) $13 + 22 =$ ______

2) $71 + 23 =$ ______

3) $61 + 58 =$ ______

4) $56 + 85 =$ ______

5) $368 + 249 =$ ______

6) $330 + 903 =$ ______

7) $471 + 293 =$ ______

8) $1{,}950 + 2{,}655 =$ ______

Estimate the product by rounding each number to the nearest ten.

9) $32 \times 71 =$ ______

10) $12 \times 33 =$ ______

11) $31 \times 83 =$ ______

12) $19 \times 11 =$ ______

13) $42 \times 76 =$ ______

14) $63 \times 34 =$ ______

15) $19 \times 31 =$ ______

16) $59 \times 71 =$ ______

Estimate the sum or product by rounding each number to the nearest ten.

17) $29 \times 12 =$ ______

18) $37 \times 26 =$ ______

19) $48 + 82 =$ ______

20) $65 + 44 =$ ______

21) $37 \times 14 =$ ______

22) $71 + 32 =$ ______

Adding and Subtracting Integers

Find each sum.

1) $14 + (-6) =$

2) $(-13) + (-20) =$

3) $5 + (-28) =$

4) $50 + (-12) =$

5) $(-7) + (-15) + 3 =$

6) $30 + (-14) + 8 =$

7) $40 + (-10) + (-14) + 17 =$

8) $(-15) + (-20) + 13 + 35 =$

9) $40 + (-20) + (38 - 29) =$

10) $28 + (-12) + (30 - 12) =$

Find each difference.

11) $(-18) - (-7) =$

12) $25 - (-14) =$

13) $(-20) - 36 =$

14) $34 - (-19) =$

15) $51 - (30 - 21) =$

16) $17 - (5) - (-24) =$

17) $(35 + 20) - (-46) =$

18) $48 - 16 - (-8) =$

19) $62 - (28 + 17) - (-15) =$

20) $58 - (-23) - (-31) =$

21) $19 - (-8) - (-13) =$

22) $(19 - 24) - (-14) =$

23) $27 - 33 - (-21) =$

24) $58 - (32 + 24) - (-9) =$

25) $36 - (-30) + (-17) =$

26) $27 - (-42) + (-31) =$

Multiplying and Dividing Integers

Find each product.

1) $(-9) \times (-5) =$

2) $(-3) \times 9 =$

3) $8 \times (-12) =$

4) $(-7) \times (-20) =$

5) $(-3) \times (-5) \times 6 =$

6) $(14 - 3) \times (-8) =$

7) $12 \times (-9) \times (-3) =$

8) $(140 + 10) \times (-2) =$

9) $10 \times (-12 + 8) \times 3 =$

10) $(-8) \times (-5) \times (-10) =$

Find each quotient.

11) $42 \div (-7) =$

12) $(-48) \div (-6) =$

13) $(-40) \div (-8) =$

14) $54 \div (-2) =$

15) $152 \div 19 =$

16) $(-144) \div (-12) =$

17) $180 \div (-10) =$

18) $(-312) \div (-12) =$

19) $221 \div (-13) =$

20) $(-126) \div (6) =$

21) $(-161) \div (-7) =$

22) $-266 \div (-14) =$

23) $(-120) \div (-4) =$

24) $270 \div (-18) =$

25) $(-208) \div (-8) =$

26) $(135) \div (-15) =$

Order of Operations

Evaluate each expression.

1) $7 + (5 \times 4) =$

2) $14 - (3 \times 6) =$

3) $(19 \times 4) + 16 =$

4) $(16 - 7) - (8 \times 2) =$

5) $27 + (18 \div 3) =$

6) $(18 \times 8) \div 6 =$

7) $(32 \div 4) \times (-2) =$

8) $(9 \times 4) + (32 - 18) =$

9) $24 + (4 \times 3) + 7 =$

10) $(36 \times 3) \div (2 + 2) =$

11) $(-7) + (12 \times 3) + 11 =$

12) $(8 \times 5) - (24 \div 6) =$

13) $(7 \times 6 \div 3) - (12 + 9) =$

14) $(13 + 5 - 14) \times 3 - 2 =$

15) $(20 - 14 + 30) \times (64 \div 4) =$

16) $32 + \big(28 - (36 \div 9)\big) =$

17) $(7 + 6 - 4 - 7) + (15 \div 5) =$

18) $(85 - 20) + (20 - 18 + 7) =$

19) $(20 \times 2) + (14 \times 3) - 22 =$

20) $18 + 5 - (30 \times 3) + 20 =$

Ordering Integers and Numbers

Order each set of integers from least to greatest.

1) $8, -10, -5, -3, 4$ ___, ___, ___, ___, ___, ___

2) $-10, -18, 6, 14, 27$ ___, ___, ___, ___, ___, ___

3) $15, -8, -21, 21, -23$ ___, ___, ___, ___, ___, ___

4) $-14, -40, 23, -12, 47$ ___, ___, ___, ___, ___, ___

5) $59, -54, 32, -57, 36$ ___, ___, ___, ___, ___, ___

6) $68, 26, -19, 47, -34$ ___, ___, ___, ___, ___, ___

Order each set of integers from greatest to least.

7) $18, 36, -16, -18, -10$ ___, ___, ___, ___, ___, ___

8) $27, 34, -12, -24, 94$ ___, ___, ___, ___, ___, ___

9) $50, -21, -13, 42, -2$ ___, ___, ___, ___, ___, ___

10) $37, 46, -20, -16, 86$ ___, ___, ___, ___, ___, ___

11) $-18, 88, -26, -59, 75$ ___, ___, ___, ___, ___, ___

12) $-65, -30, -25, 3, 14$ ___, ___, ___, ___, ___, ___

Integers and Absolute Value

Write absolute value of each number.

1) $|-2| =$

2) $|-27| =$

3) $|-20| =$

4) $|14| =$

5) $|6| =$

6) $|-55| =$

7) $|16| =$

8) $|2| =$

9) $|54| =$

10) $|-4| =$

11) $|-11|$

12) $|88| =$

13) $|0| =$

14) $|79| =$

15) $|-32| =$

16) $|-17| =$

17) $|42| =$

18) $|-46| =$

19) $|1| =$

20) $|-40| =$

Evaluate the value.

21) $|-5| - \frac{|-21|}{7} =$

22) $14 - |3 - 15| - |-4| =$

23) $\frac{|-32|}{4} \times |-4| =$

24) $\frac{|7 \times (-3)|}{7} \times \frac{|-19|}{3} =$

25) $|4 \times (-5)| + \frac{|-40|}{5} =$

26) $\frac{|-45|}{9} \times \frac{|-24|}{12} =$

27) $|-12 + 8| \times \frac{|-7 \times 7|}{7}$

28) $\frac{|-11 \times 2|}{4} \times |-16| =$

Factoring Numbers

List all positive factors of each number.

1) 9

2) 16

3) 24

4) 30

5) 26

6) 46

7) 20

8) 68

9) 28

10) 98

11) 14

12) 54

13) 55

14) 18

15) 63

16) 34

17) 50

18) 62

19) 95

20) 64

21) 70

22) 45

23) 22

24) 65

Greatest Common Factor

Find the GCF for each number pair.

1) 6, 2

2) 4, 5

3) 3, 12

4) 7, 3

5) 5, 10

6) 8, 48

7) 6, 18

8) 9, 15

9) 12, 18

10) 4, 36

11) 6, 10

12) 28, 52

13) 25, 10

14) 22, 24

15) 9, 54

16) 8, 54

17) 42, 14

18) 16, 40

19) 9,2, 3

20) 5, 15, 10

21) 7, 9, 2

22) 16, 64

23) 30, 48

24) 36, 63

Least Common Multiple

Find the LCM for each number pair.

1) 6, 9
2) 15, 45
3) 16, 40
4) 12, 36
5) 18, 27
6) 14, 42
7) 6, 30
8) 8, 56
9) 7, 21
10) 8, 20
11) 15, 25
12) 7, 9
13) 4, 11
14) 8, 28
15) 28, 56
16) 40, 50
17) 12, 13
18) 22, 11
19) 36, 20
20) 15, 35
21) 18, 81
22) 30, 54
23) 18,45
24) 75, 25

Answers of Worksheets

Rounding

1) 40
2) 90
3) 20
4) 60
5) 20
6) 30
7) 90
8) 70
9) 50
10) 80
11) 60
12) 90
13) 200
14) 400
15) 800
16) 100
17) 300
18) 400
19) 600
20) 800
21) 600
22) 900
23) 500
24) 300
25) 1,000
26) 3,000
27) 4,000
28) 7,000
29) 9,000
30) 23,000
31) 45,000
32) 17,000
33) 53,000
34) 85,000
35) 71,000
36) 27,000

Rounding and Estimates

1) 30
2) 90
3) 120
4) 150
5) 620
6) 1,230
7) 760
8) 4,610
9) 2,100
10) 300
11) 2,400
12) 200
13) 3,200
14) 1,800
15) 600
16) 4,200
17) 300
18) 1,200
19) 130
20) 110
21) 400
22) 100

Adding and Subtracting Integers

1) 8
2) −33
3) −23
4) 38
5) −19
6) 24
7) 33
8) 13
9) 29
10) 34
11) −11
12) 39
13) −56
14) 53
15) 42
16) 36
17) 101
18) 40
19) 32
20) 112
21) 40
22) 9
23) 15
24) 11
25) 49
26) 38

Multiplying and Dividing Integers

1) 45
2) −27
3) −96
4) 140
5) 90
6) −88
7) 324
8) −300
9) −120
10) −400
11) −6
12) 8
13) 5
14) −27
15) 8
16) 12
17) −18
18) 26
19) −17
20) −21

21) 23	23) 30	25) 26
22) 19	24) −15	26) −9

Order of Operations

1) 27	6) 24	11) 40	16) 56
2) −4	7) −16	12) 36	17) 5
3) 92	8) 50	13) −7	18) 74
4) −7	9) 43	14) 10	19) 60
5) 33	10) 27	15) 576	20) −47

Ordering Integers and Numbers

1) −10, −5, −3, 4, 8	7) 36, 18, −10, −16, −18
2) −18, −10, 6, 14, 27	8) 94, 34, 27, −12, −24
3) −23, −21, −8, 15, 21	9) 50, 42, −2, −13, −21
4) −40, −14, −12, 23, 47	10) 86, 46, 37, −16, −20
5) −57, −54, 32, 36, 59	11) 88, 75, −18, −26, −59
6) −34, −19, 26, 47, 68	12) 14, 3, −25, −30, −65

Integers and Absolute Value

1) 2	8) 2	15) 32	22) −2
2) 27	9) 54	16) 17	23) 32
3) 20	10) 4	17) 42	24) 19
4) 14	11) 11	18) 46	25) 28
5) 6	12) 88	19) 1	26) 10
6) 55	13) 0	20) 40	27) 28
7) 16	14) 79	21) 2	28) 88

Factoring Numbers

1) 1, 3, 9	9) 1, 2, 4, 7, 14, 28	17) 1, 2, 5, 10, 25, 50
2) 1, 2, 4, 8, 16	10) 1, 2, 7, 14, 49, 98	18) 1, 2, 31, 62
3) 1, 2, 3, 4, 6, 8, 12, 24	11) 1, 2, 7, 14	19) 1, 5, 19, 95
4) 1, 2, 3, 5, 6, 10, 15, 30	12) 1, 2, 3, 6, 9, 18, 27, 54	20) 1, 2, 4, 8, 16, 32, 64
5) 1, 2, 13, 26	13) 1, 5,11,55	21) 1, 2, 5, 7, 10, 14, 35, 70
6) 1, 2, 23, 46	14) 1, 2, 3, 6, 9, 18	22) 1, 3, 5, 9, 15, 45
7) 1, 2, 4, 5, 10, 20	15) 1, 3, 7, 9, 21, 63	23) 1, 2, 11, 22
8) 1, 2, 4, 17, 34, 68	16) 1, 2, 17, 34	24) 1,5,13, 65

Greatest Common Factor

1) 2
2) 1
3) 3
4) 1
5) 5
6) 8
7) 6
8) 3
9) 6
10) 4
11) 2
12) 4
13) 5
14) 2
15) 9
16) 2
17) 14
18) 8
19) 1
20) 5
21) 1
22) 16
23) 6
24) 9

Least Common Multiple

1) 18
2) 45
3) 80
4) 36
5) 54
6) 42
7) 30
8) 56
9) 21
10) 40
11) 75
12) 63
13) 44
14) 56
15) 56
16) 200
17) 156
18) 22
19) 180
20) 105
21) 162
22) 270
23) 90
24) 75

Chapter 2 :
Fractions and Decimals

Topics that you'll practice in this chapter:

- ✓ Simplifying Fractions
- ✓ Adding and Subtracting Fractions
- ✓ Multiplying and Dividing Fractions
- ✓ Adding and Subtract Mixed Numbers
- ✓ Multiplying and Dividing Mixed Numbers
- ✓ Adding and Subtracting Decimals
- ✓ Multiplying and Dividing Decimals
- ✓ Comparing Decimals
- ✓ Rounding Decimals

"A Man is like a fraction whose numerator is what he is and whose denominator is what he thinks of himself. The larger the denominator, the smaller the fraction." –Tolstoy

Simplifying Fractions

Simplify each fraction to its lowest terms.

1) $\frac{5}{10} =$

2) $\frac{28}{35} =$

3) $\frac{27}{36} =$

4) $\frac{40}{80} =$

5) $\frac{14}{56} =$

6) $\frac{32}{48} =$

7) $\frac{52}{65} =$

8) $\frac{15}{60} =$

9) $\frac{80}{160} =$

10) $\frac{55}{77} =$

11) $\frac{28}{112} =$

12) $\frac{32}{64} =$

13) $\frac{63}{72} =$

14) $\frac{81}{90} =$

15) $\frac{35}{105} =$

16) $\frac{25}{70} =$

17) $\frac{80}{280} =$

18) $\frac{12}{81} =$

19) $\frac{36}{186} =$

20) $\frac{240}{540} =$

21) $\frac{70}{560} =$

Find the answer for each problem.

22) Which of the following fractions equal to $\frac{3}{4}$? ____

A. $\frac{60}{90}$ B. $\frac{43}{104}$ C. $\frac{48}{64}$ D. $\frac{150}{300}$

23) Which of the following fractions equal to $\frac{5}{8}$? ____

A. $\frac{125}{200}$ B. $\frac{115}{200}$ C. $\frac{50}{100}$ D. $\frac{30}{90}$

24) Which of the following fractions equal to $\frac{3}{7}$? ____

A. $\frac{58}{116}$ B. $\frac{54}{126}$ C. $\frac{270}{167}$ D. $\frac{42}{63}$

Adding and Subtracting Fractions

Find the sum.

1) $\frac{5}{9}+\frac{4}{9}=$

2) $\frac{1}{2}+\frac{1}{7}=$

3) $\frac{3}{8}+\frac{1}{4}=$

4) $\frac{3}{5}+\frac{1}{2}=$

5) $\frac{1}{4}+\frac{3}{5}=$

6) $\frac{7}{8}+\frac{3}{8}=$

7) $\frac{1}{2}+\frac{7}{10}=$

8) $\frac{2}{5}+\frac{2}{3}=$

9) $\frac{5}{7}+\frac{2}{3}=$

10) $\frac{7}{12}+\frac{3}{4}=$

11) $\frac{5}{6}+\frac{2}{5}=$

12) $\frac{1}{12}+\frac{2}{3}=$

Find the difference.

13) $\frac{1}{3}-\frac{1}{6}=$

14) $\frac{3}{4}-\frac{1}{8}=$

15) $\frac{1}{2}-\frac{1}{3}=$

16) $\frac{1}{4}-\frac{1}{5}=$

17) $\frac{5}{8}-\frac{2}{3}=$

18) $\frac{1}{4}-\frac{1}{7}=$

19) $\frac{5}{6}-\frac{1}{9}=$

20) $\frac{3}{4}-\frac{1}{6}=$

21) $\frac{7}{8}-\frac{1}{12}=$

22) $\frac{8}{15}-\frac{3}{5}=$

23) $\frac{3}{12}-\frac{1}{14}=$

24) $\frac{10}{13}-\frac{7}{26}=$

25) $\frac{6}{7}-\frac{3}{4}=$

26) $\frac{4}{5}-\frac{1}{8}=$

27) $\frac{4}{7}-\frac{2}{35}=$

28) $\frac{9}{16}-\frac{2}{8}=$

29) $\frac{8}{9}-\frac{7}{18}=$

30) $\frac{1}{2}-\frac{4}{9}=$

Multiplying and Dividing Fractions

Find the value of each expression in lowest terms.

1) $\frac{1}{5} \times \frac{15}{5} =$

2) $\frac{9}{12} \times \frac{4}{9} =$

3) $\frac{1}{16} \times \frac{8}{10} =$

4) $\frac{1}{24} \times \frac{8}{10} =$

5) $\frac{1}{5} \times \frac{1}{4} =$

6) $\frac{7}{9} \times \frac{1}{7} =$

7) $\frac{6}{7} \times \frac{1}{3} =$

8) $\frac{2}{8} \times \frac{2}{8} =$

9) $\frac{5}{8} \times \frac{3}{5} =$

10) $\frac{4}{7} \times \frac{1}{8} =$

11) $\frac{7}{15} \times \frac{5}{7} =$

12) $\frac{3}{10} \times \frac{5}{9} =$

Find the value of each expression in lowest terms.

13) $\frac{1}{4} \div \frac{1}{8} =$

14) $\frac{1}{10} \div \frac{1}{5} =$

15) $\frac{3}{4} \div \frac{1}{5} =$

16) $\frac{1}{3} \div \frac{5}{6} =$

17) $\frac{1}{7} \div \frac{8}{42} =$

18) $\frac{3}{4} \div \frac{1}{6} =$

19) $\frac{2}{7} \div \frac{7}{13} =$

20) $\frac{1}{24} \div \frac{3}{16} =$

21) $\frac{7}{12} \div \frac{5}{6} =$

22) $\frac{22}{18} \div \frac{11}{9} =$

23) $\frac{9}{35} \div \frac{3}{7} =$

24) $\frac{2}{7} \div \frac{8}{21} =$

25) $\frac{1}{9} \div \frac{2}{5} =$

26) $\frac{5}{12} \div \frac{3}{5} =$

27) $\frac{3}{20} \div \frac{1}{6} =$

28) $\frac{8}{20} \div \frac{3}{4} =$

29) $\frac{5}{6} \div \frac{2}{9} =$

30) $\frac{5}{11} \div \frac{3}{4} =$

Adding and Subtracting Mixed Numbers

Find the sum.

1) $3\frac{1}{3}+2\frac{1}{6}=$

2) $4\frac{1}{2}+3\frac{1}{2}=$

3) $3\frac{3}{8}+1\frac{1}{8}=$

4) $2\frac{1}{4}+2\frac{1}{3}=$

5) $3\frac{5}{6}+2\frac{7}{12}=$

6) $5\frac{4}{15}+3\frac{3}{5}=$

7) $2\frac{1}{3}+4\frac{3}{7}=$

8) $3\frac{1}{2}+4\frac{2}{5}=$

9) $5\frac{2}{5}+6\frac{3}{7}=$

10) $8\frac{5}{16}+6\frac{1}{12}=$

Find the difference.

11) $3\frac{1}{4}-1\frac{3}{4}=$

12) $6\frac{3}{5}-4\frac{2}{5}=$

13) $4\frac{1}{3}-3\frac{1}{9}=$

14) $7\frac{1}{7}-5\frac{1}{2}=$

15) $5\frac{1}{3}-2\frac{1}{12}=$

16) $8\frac{1}{5}-4\frac{1}{3}=$

17) $9\frac{1}{4}-6\frac{1}{8}=$

18) $11\frac{7}{15}-8\frac{3}{5}=$

19) $14\frac{5}{6}-11\frac{3}{5}=$

20) $18\frac{2}{7}-14\frac{1}{5}=$

21) $9\frac{1}{3}-4\frac{1}{4}=$

22) $6\frac{1}{8}-4\frac{1}{16}=$

23) $19\frac{3}{8}-15\frac{1}{3}=$

24) $11\frac{1}{9}-8\frac{1}{8}=$

25) $17\frac{1}{7}-11\frac{1}{5}=$

26) $16\frac{2}{9}-9\frac{5}{7}=$

Multiplying and Dividing Mixed Numbers

Find the product.

1) $5\frac{1}{2} \times 2\frac{1}{4} =$

2) $5\frac{1}{3} \times 4\frac{1}{3} =$

3) $5\frac{3}{4} \times 6\frac{1}{4} =$

4) $3\frac{1}{3} \times 2\frac{3}{5} =$

5) $4\frac{8}{10} \times 1\frac{1}{24} =$

6) $6\frac{2}{7} \times 1\frac{1}{11} =$

7) $8\frac{2}{3} \times 3\frac{1}{2} =$

8) $3\frac{4}{7} \times 2\frac{1}{5} =$

9) $5\frac{2}{8} \times 4\frac{1}{6} =$

10) $7\frac{3}{3} \times 1\frac{3}{8} =$

Find the quotient.

11) $2\frac{2}{5} \div 4\frac{1}{5} =$

12) $4\frac{1}{6} \div 3\frac{1}{3} =$

13) $6\frac{1}{3} \div 1\frac{1}{2} =$

14) $7\frac{1}{10} \div 2\frac{2}{5} =$

15) $3\frac{1}{3} \div 1\frac{1}{9} =$

16) $1\frac{1}{10} \div 4\frac{1}{2} =$

17) $1\frac{3}{16} \div 5\frac{1}{4} =$

18) $4\frac{1}{3} \div 4\frac{3}{4} =$

19) $9\frac{1}{3} \div 2\frac{1}{4} =$

20) $15\frac{1}{3} \div 5\frac{1}{2} =$

21) $4\frac{1}{6} \div 1\frac{1}{5} =$

22) $1\frac{1}{18} \div 1\frac{2}{9} =$

23) $4\frac{2}{7} \div 1\frac{3}{10} =$

24) $7\frac{1}{3} \div 2\frac{2}{11} =$

25) $8\frac{2}{5} \div 1\frac{1}{6} =$

26) $9\frac{1}{3} \div 2\frac{1}{7} =$

Adding and Subtracting Decimals

Add and subtract decimals.

1) $\begin{array}{r} 35.19 \\ -\ 24.28 \\ \hline \end{array}$ ____

4) $\begin{array}{r} 38.72 \\ -\ 21.68 \\ \hline \end{array}$ ____

7) $\begin{array}{r} 86.09 \\ -\ 35.14 \\ \hline \end{array}$ ____

2) $\begin{array}{r} 34.29 \\ +\ 42.58 \\ \hline \end{array}$ ____

5) $\begin{array}{r} 57.39 \\ +\ 26.54 \\ \hline \end{array}$ ____

8) $\begin{array}{r} 54.51 \\ +\ 32.66 \\ \hline \end{array}$ ____

3) $\begin{array}{r} 61.20 \\ +\ 33.75 \\ \hline \end{array}$ ____

6) $\begin{array}{r} 70.24 \\ -\ 42.35 \\ \hline \end{array}$ ____

9) $\begin{array}{r} 114.21 \\ -\ 88.69 \\ \hline \end{array}$ ____

Find the missing number.

10) ___ + 2.8 = 5.4

11) 4.1 + ___ = 5.88

12) 6.45 + ___ = 8

13) 7.25 − ___ = 3.40

14) ___ − 2.35 = 4.25

15) ___ − 19.85 = 6.54

16) 22.15 + ___ = 28.95

17) ___ − 37.16 = 9.42

18) ___ + 24.50 = 34.19

19) 72.40 + ___ = 125.20

Multiplying and Dividing Decimals

Find the product.

1) $0.5 \times 0.6 =$

2) $3.3 \times 0.4 =$

3) $1.28 \times 0.5 =$

4) $0.35 \times 0.6 =$

5) $1.85 \times 0.6 =$

6) $0.24 \times 0.5 =$

7) $5.25 \times 1.4 =$

8) $18.5 \times 4.6 =$

9) $15.4 \times 6.8 =$

10) $19.5 \times 2.6 =$

11) $32.2 \times 1.5 =$

12) $78.4 \times 4.5 =$

Find the quotient.

13) $1.85 \div 10 =$

14) $74.6 \div 100 =$

15) $3.6 \div 3 =$

16) $9.6 \div 0.4 =$

17) $15.5 \div 0.5 =$

18) $32.8 \div 0.2 =$

19) $22.15 \div 1{,}000 =$

20) $53.55 \div 0.7 =$

21) $322.2 \div 0.2 =$

22) $50.67 \div 0.18 =$

23) $77.4 \div 0.8 =$

24) $27.93 \div 0.03 =$

Comparing Decimals

Write the correct comparison symbol (>, < or =).

1) 0.70 □ 0.070

2)0.049 □ 0.49

3)5.090 □ 5.09

4)2.57 □ 2.05

5)9.03 □ 0.930

6)6.06 □ 6.6

7)7.02 □ 7.020

8)3.04 □ 3.2

9)3.61 □ 3.245

10) 0.986 □ 0.0986

11) 17.24 □ 17.240

12) 0.759 □ 0.81

13) 9.040 □9.40

14) 5.73 □ 5.213

15) 9.44 □ 9.404

16) 7.17 □ 7.170

17) 4.85 □ 4.085

18) 9.041 □ 9.40

19) 3.033 □ 3.030

20) 4.97 □ 4.970

Rounding Decimals

Round each decimal to the nearest whole number.

1) 28.12
2) 6.9
3) 16.22
4) 8.5
5) 7.95
6) 52.7

Round each decimal to the nearest tenth.

7) 31.761
8) 14.421
9) 94.729
10) 77.89
11) 13.219
12) 59.89

Round each decimal to the nearest hundredth.

13) 8.428
14) 23.812
15) 55.3786
16) 231.912
17) 62.241
18) 19.447

Round each decimal to the nearest thousandth.

19) 15.54324
20) 34.62586
21) 243.8652
22) 80.4529
23) 67.1983
24) 72.36788

Answers of Worksheets

Simplifying Fractions

1) $\frac{1}{2}$
2) $\frac{4}{5}$
3) $\frac{3}{4}$
4) $\frac{1}{2}$
5) $\frac{1}{4}$
6) $\frac{2}{3}$
7) $\frac{4}{5}$
8) $\frac{1}{4}$
9) $\frac{1}{2}$
10) $\frac{5}{7}$
11) $\frac{1}{4}$
12) $\frac{1}{2}$
13) $\frac{7}{8}$
14) $\frac{9}{10}$
15) $\frac{1}{3}$
16) $\frac{5}{14}$
17) $\frac{2}{7}$
18) $\frac{4}{27}$
19) $\frac{6}{31}$
20) $\frac{4}{9}$
21) $\frac{1}{8}$
22) C
23) A
24) B

Adding and Subtracting Fractions

1) $\frac{9}{9} = 1$
2) $\frac{9}{14}$
3) $\frac{5}{8}$
4) $1\frac{1}{10}$
5) $\frac{17}{20}$
6) $1\frac{1}{4}$
7) $1\frac{1}{5}$
8) $1\frac{1}{15}$
9) $1\frac{8}{21}$
10) $1\frac{1}{3}$
11) $1\frac{7}{30}$
12) $\frac{3}{4}$
13) $\frac{1}{6}$
14) $\frac{5}{8}$
15) $\frac{1}{6}$
16) $\frac{1}{20}$
17) $-\frac{1}{24}$
18) $\frac{3}{28}$
19) $\frac{13}{18}$
20) $\frac{7}{12}$
21) $\frac{19}{24}$
22) $-\frac{1}{15}$
23) $\frac{5}{28}$
24) $\frac{1}{2}$
25) $\frac{3}{28}$
26) $\frac{27}{40}$
27) $\frac{18}{35}$
28) $\frac{5}{16}$
29) $\frac{1}{2}$
30) $\frac{1}{18}$

Multiplying and Dividing Fractions

1) $\frac{3}{5}$
2) $\frac{1}{3}$
3) $\frac{1}{20}$
4) $\frac{1}{30}$
5) $\frac{1}{20}$
6) $\frac{1}{9}$
7) $\frac{2}{7}$
8) $\frac{1}{16}$
9) $\frac{3}{8}$
10) $\frac{1}{14}$
11) $\frac{1}{3}$
12) $\frac{1}{6}$
13) 2
14) $\frac{1}{2}$
15) $3\frac{3}{4}$
16) $\frac{2}{5}$

17) $\frac{3}{4}$

18) $4\frac{1}{2}$

19) $\frac{26}{49}$

20) $\frac{2}{9}$

21) $\frac{7}{10}$

22) 1

23) $\frac{3}{5}$

24) $\frac{3}{4}$

25) $\frac{5}{18}$

26) $\frac{25}{36}$

27) $\frac{9}{10}$

28) $\frac{8}{15}$

29) $3\frac{3}{4}$

30) $\frac{20}{33}$

Adding and Subtracting Mixed Numbers

1) $5\frac{1}{2}$

2) 8

3) $4\frac{1}{2}$

4) $4\frac{7}{12}$

5) $6\frac{5}{12}$

6) $8\frac{13}{15}$

7) $6\frac{16}{21}$

8) $7\frac{9}{10}$

9) $11\frac{29}{35}$

10) $14\frac{19}{48}$

11) $1\frac{1}{2}$

12) $2\frac{1}{5}$

13) $1\frac{2}{9}$

14) $1\frac{9}{14}$

15) $3\frac{1}{4}$

16) $3\frac{13}{15}$

17) $3\frac{1}{8}$

18) $2\frac{13}{15}$

19) $3\frac{7}{30}$

20) $4\frac{3}{35}$

21) $5\frac{1}{12}$

22) $2\frac{1}{16}$

23) $4\frac{1}{24}$

24) $2\frac{71}{72}$

25) $5\frac{33}{35}$

26) $6\frac{32}{63}$

Multiplying and Dividing Mixed Numbers

1) $12\frac{3}{8}$

2) $23\frac{1}{9}$

3) $35\frac{15}{16}$

4) $8\frac{2}{3}$

5) 5

6) $6\frac{6}{7}$

7) $30\frac{1}{3}$

8) $7\frac{6}{7}$

9) $21\frac{7}{8}$

10) 11

11) $\frac{4}{7}$

12) $1\frac{1}{4}$

13) $4\frac{2}{9}$

14) $2\frac{23}{24}$

15) 3

16) $\frac{11}{45}$

17) $\frac{19}{84}$

18) $\frac{52}{57}$

19) $4\frac{4}{27}$

20) $2\frac{26}{33}$

21) $3\frac{17}{36}$

22) $\frac{19}{22}$

23) $3\frac{27}{91}$

24) $3\frac{13}{36}$

25) $7\frac{1}{5}$

26) $4\frac{16}{45}$

Adding and Subtracting Decimals

1) 10.91

2) 76.87

3) 94.95

4) 17.04

5) 83.93
6) 27.89
7) 50.95
8) 87.17
9) 25.52
10) 2.6
11) 1.78
12) 1.55
13) 3.85
14) 6.6
15) 26.39
16) 6.8
17) 46.58
18) 9.69
19) 52.8

Multiplying and Dividing Decimals

1) 0.3
2) 1.32
3) 0.64
4) 0.21
5) 1.11
6) 0.12
7) 7.35
8) 85.1
9) 104.72
10) 50.7
11) 48.3
12) 352.8
13) 0.185
14) 0.746
15) 1.2
16) 24
17) 31
18) 164
19) 0.02215
20) 76.5
21) 1,611
22) 281.5
23) 96.75
24) 931

Comparing Decimals

1) >
2) <
3) =
4) >
5) >
6) <
7) =
8) <
9) >
10) >
11) =
12) <
13) <
14) >
15) >
16) =
17) >
18) <
19) >
20) =

Rounding Decimals

1) 28
2) 7
3) 16
4) 9
5) 8
6) 53
7) 31.8
8) 14.4
9) 94.7
10) 77.9
11) 13.2
12) 59.9
13) 8.43
14) 23.81
15) 55.38
16) 231.91
17) 62.24
18) 19.45
19) 15.543
20) 34.626
21) 243.865
22) 80.453
23) 67.198
24) 72.368

Chapter 3 :

Proportions, Ratios, and Percent

Topics that you'll practice in this chapter:

- ✓ Simplifying Ratios
- ✓ Proportional Ratios
- ✓ Similarity and Ratios
- ✓ Ratio and Rates Word Problems
- ✓ Percentage Calculations
- ✓ Percent Problems
- ✓ Discount, Tax and Tip
- ✓ Percent of Change
- ✓ Simple Interest

Without mathematics, there's nothing you can do. Everything around you is mathematics. Everything around you is numbers." – Shakuntala Devi

Simplifying Ratios

Reduce each ratio.

1) 15: 20 = ___:___
2) 7: 70 = ___:___
3) 16: 28 = ___:___
4) 7: 21 = ___:___
5) 4: 40 = ___:___
6) 6: 48 = ___:___
7) 16: 64 = ___:___
8) 10: 25 = ___:___
9) 8: 48 = ___:___
10) 49: 63 = ___:___
11) 18: 27 = ___:___
12) 35: 10 = ___:___
13) 90: 9 = ___:___
14) 24: 32 = ___:___
15) 7: 56 = ___:___
16) 45: 63 = ___:___
17) 56: 72 = ___:___
18) 26: 13 = ___:___
19) 15: 45 = ___:___
20) 28: 4 = ___:___
21) 24: 48 = ___:___
22) 30: 24 = ___:___
23) 70: 140 = ___:___
24) 6: 180 = ___:___

Write each ratio as a fraction in simplest form.

25) 6: 12 =
26) 30: 50 =
27) 15: 35 =
28) 9: 27 =
29) 8: 24 =
30) 18: 84 =
31) 7: 14 =
32) 7: 35 =
33) 40: 96 =
34) 12: 54 =
35) 44: 52 =
36) 12: 27 =
37) 15: 180 =
38) 39: 143 =
39) 20: 300 =
40) 30: 120 =
41) 56: 42 =
42) 26: 130 =
43) 66: 123 =
44) 70: 630 =
45) 75: 125 =

Proportional Ratios

Fill in the blanks; Calculate each proportion.

1) $3:8 = \_\_ : 48$

2) $2:5 = 20: \_\_$

3) $1:9 = \_\_ : 81$

4) $6:7 = 12: \_\_$

5) $9:2 = 63: \_\_$

6) $8:7 = \_\_ : 49$

7) $20:3 = \_\_ : 15$

8) $1:3 = \_\_ : 75$

9) $7:6 = \_\_ : 60$

10) $8:5 = \_\_ : 45$

11) $3:10 = 60: \_\_$

12) $6:11 = 42: \_\_$

State if each pair of ratios form a proportion.

13) $\frac{3}{20}$ *and* $\frac{9}{60}$

14) $\frac{1}{7}$ *and* $\frac{6}{42}$

15) $\frac{3}{7}$ *and* $\frac{24}{56}$

16) $\frac{4}{9}$ *and* $\frac{12}{18}$

17) $\frac{1}{9}$ *and* $\frac{12}{81}$

18) $\frac{7}{8}$ *and* $\frac{21}{28}$

19) $\frac{9}{13}$ *and* $\frac{27}{39}$

20) $\frac{1}{8}$ *and* $\frac{8}{64}$

21) $\frac{6}{19}$ *and* $\frac{30}{85}$

22) $\frac{5}{9}$ *and* $\frac{40}{81}$

23) $\frac{9}{14}$ *and* $\frac{108}{168}$

24) $\frac{15}{23}$ *and* $\frac{360}{552}$

Calculate each proportion.

25) $\frac{20}{25} = \frac{32}{x}, x = \_\_\_\_$

26) $\frac{1}{8} = \frac{32}{x}, x = \_\_\_\_$

27) $\frac{15}{5} = \frac{21}{x}, x = \_\_\_\_$

28) $\frac{1}{7} = \frac{x}{294}, x = \_\_\_\_$

29) $\frac{7}{9} = \frac{x}{81}, x = \_\_\_\_$

30) $\frac{1}{5} = \frac{13}{x}, x = \_\_\_\_$

31) $\frac{9}{5} = \frac{36}{x}, x = \_\_\_\_$

32) $\frac{6}{13} = \frac{48}{x}, x = \_\_\_\_$

33) $\frac{5}{8} = \frac{x}{88}, x = \_\_\_\_$

34) $\frac{4}{15} = \frac{x}{240}, x = \_\_\_\_$

35) $\frac{9}{19} = \frac{x}{266}, x = \_\_\_\_$

36) $\frac{7}{15} = \frac{x}{270}, x = \_\_\_\_$

Similarity and Ratios

Each pair of figures is similar. Find the missing side.

1)

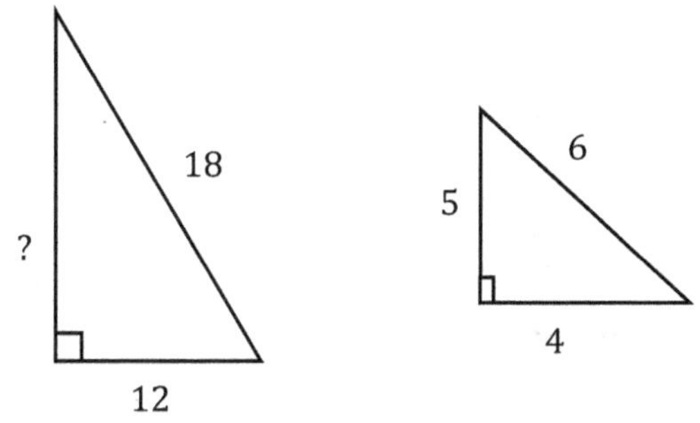

2)

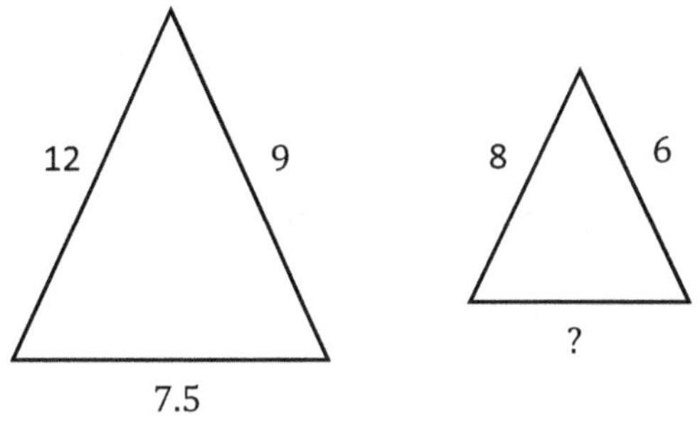

3)

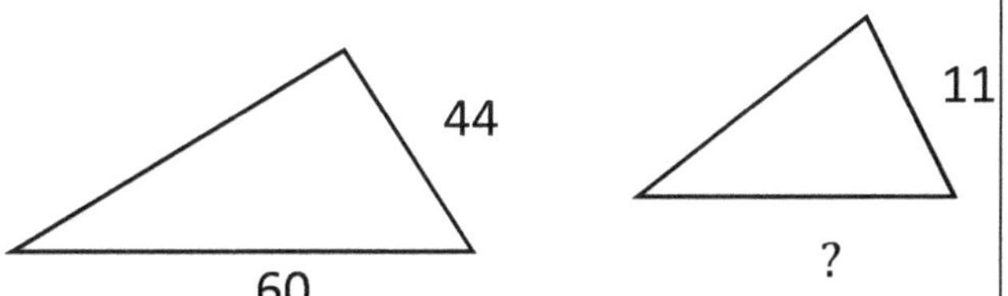

4)

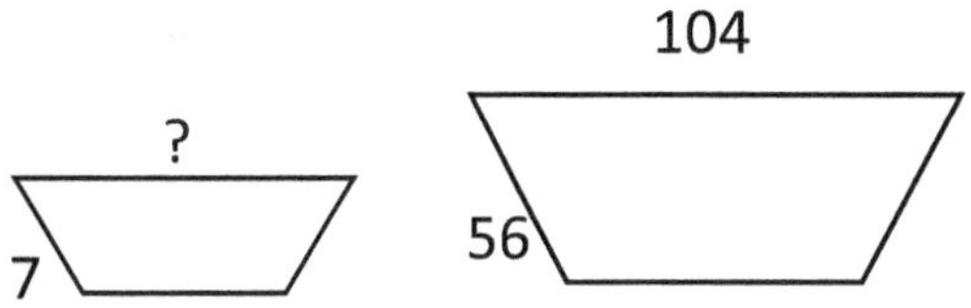

Calculate.

5) Two rectangles are similar. The first is 24 feet wide and 120 feet long. The second is 30 feet wide. What is the length of the second rectangle? ________________

6) Two rectangles are similar. One is 5 meters by 36 meters. The longer side of the second rectangle is 90 meters. What is the other side of the second rectangle? ________________

7) A building casts a shadow 25 ft long. At the same time a girl 10 ft tall casts a shadow 5 ft long. How tall is the building? ________________

8) The scale of a map of Texas is 4 inches: 32 miles. If you measure the distance from Dallas to Martin County as 38.4 inches, approximately how far is Martin County from Dallas? ________

Ratio and Rates Word Problems

Find the answer for each word problem.

1) Mason has 24 red cards and 36 green cards. What is the ratio of Mason 's red cards to his green cards? ____________

2) In a party, 45 soft drinks are required for every 54 guests. If there are 378 guests, how many soft drinks is required? ____________

3) In Mason's class, 42 of the students are tall and 24 are short. In Michael's class 84 students are tall and 48 students are short. Which class has a higher ratio of tall to short students? ____________

4) The price of 5 apples at the Quick Market is $4.6. The price of 7 of the same apples at Walmart is $5.95. Which place is the better buy? ____________

5) The bakers at a Bakery can make 90 bagels in 3 hours. How many bagels can they bake in 24 hours? What is that rate per hour? ____________

6) You can buy 5 cans of green beans at a supermarket for $5.75. How much does it cost to buy 45 cans of green beans? ____________

7) The ratio of boys to girls in a class is 4: 7. If there are 32 boys in the class, how many girls are in that class? ____________

8) The ratio of red marbles to blue marbles in a bag is 3: 7. If there are 50 marbles in the bag, how many of the marbles are red? ____________

Percentage Calculations

Calculate the given percent of each value.

1) $3\%\ of\ 60 =$ ____
2) $20\%\ of\ 32 =$ ____
3) $4\%\ of\ 72 =$ ____
4) $16\%\ of\ 32 =$ ____
5) $25\%\ of\ 124 =$ ____
6) $35\%\ of\ 56 =$ ____
7) $15\%\ of\ 20 =$ ____
8) $14\%\ of\ 150 =$ ____
9) $80\%\ of\ 50 =$ ____
10) $12\%\ of\ 115 =$ ____
11) $72\%\ of\ 250 =$ ____
12) $52\%\ of\ 500 =$ ____
13) $70\%\ of\ 400 =$ ____
14) $27\%\ of\ 145 =$ ____
15) $90\%\ of\ 64 =$ ____
16) $60\%\ of\ 55 =$ ____
17) $22\%\ of\ 210 =$ ____
18) $8\%\ of\ 235 =$ ____

Calculate the percent of each given value.

19) ____$\%\ of\ 25 = 5$
20) ____$\%\ of\ 40 = 20$
21) ____$\%\ of\ 25 = 2$
22) ____$\%\ of\ 50 = 16$
23) ____$\%\ of\ 250 = 5$
24) ____$\%\ of\ 40 = 32$
25) ____$\%\ of\ 125 = 20$
26) ____$\%\ of\ 700 = 49$
27) ____$\%\ of\ 350 = 49$
28) ____% of 500 = 210

Calculate each percent problem.

29) A Cinema has 250 seats. 60 seats were sold for the current movie. What percent of seats are empty? _____ %

30) There are 68 boys and 92 girls in a class. 75% of the students in the class take the bus to school. How many students do not take the bus to school?

_____

Percent Problems

Calculate each problem.

1) 9 is what percent of 45? ____%
2) 60 is what percent of 120? ____%
3) 10 is what percent of 200? ____%
4) 15 is what percent of 125? ____%
5) 10 is what percent of 400? ____%
6) 66 is what percent of 55? ____%
7) 40 is what percent of 160? ____%
8) 40 is what percent of 50? ____%
9) 120 is what percent of 800? ____%
10) 78 is what percent of 120? ____%
11) 36 is what percent of 144? ____%
12) 17 is what percent of 85? ____%
13) 90 is what percent of 900? ____%
14) 36 is what percent of 16? ____%
15) 63 is what percent of 14? ____%
16) 18 is what percent of 60? ____%
17) 126 is what percent of 200? ____%
18) 232 is what percent of 40? ____%

Calculate each percent word problem.

19) There are 40 employees in a company. On a certain day, 25 were present. What percent showed up for work? _____%
20) A metal bar weighs 60 ounces. 25% of the bar is gold. How many ounces of gold are in the bar? ____________
21) A crew is made up of 12 women; the rest are men. If 15% of the crew are women, how many people are in the crew? ____________
22) There are 40 students in a class and 8 of them are girls. What percent are boys? _____%
23) The Royals softball team played 400 games and won 280 of them. What percent of the games did they lose? _____%

Discount, Tax and Tip

Find the selling price of each item.

1) Original price of a computer: $420
Tax: 8% Selling price: $______

2) Original price of a laptop: $280
Tax: 4% Selling price: $______

3) Original price of a sofa: $820
Tax: 5% Selling price: $______

4) Original price of a car: $15,800
Tax: 3.6% Selling price: $______

5) Original price of a Table: $250
Tax: 9% Selling price: $______

6) Original price of a house: $630,000
Tax: 1.8% Selling price: $______

7) Original price of a tablet: $450
Discount: 30% Selling price: $____

8) Original price of a chair: $390
Discount: 8% Selling price: $____

9) Original price of a book: $75
Discount: 42% Selling price: $____

10) Original price of a cellphone: $820
Discount: 23% Selling price: $___

11) Food bill: $45
Tip: 15% Price: $______

12) Food bill: $32
Tipp: 20% Price: $______

13) Food bill: $90
Tip: 35% Price: $______

14) Food bill: $42
Tipp: 12% Price: $______

Find the answer for each word problem.

15) Nicolas hired a moving company. The company charged $500 for its services, and Nicolas gives the movers a 40% tip. How much does Nicolas tip the movers? $______

16) Mason has lunch at a restaurant and the cost of his meal is $90. Mason wants to leave a 25% tip. What is Mason's total bill including tip? $______

17) The sales tax in Texas is 19.80% and an item costs $350. How much is the tax? $______

18) The price of a table at Best Buy is $680. If the sales tax is 5%, what is the final price of the table including tax? $______

Percent of Change

Find each percent of change.

1) From 150 to 450. ___ %

2) From 50 ft to 250 ft. ___ %

3) From $60 to $360. ___ %

4) From 60 cm to 180 cm. ___ %

5) From 15 to 45. ___ %

6) From 80 to 16. ___ %

7) From 120 to 360. ___ %

8) From 900 to 450. ___ %

9) From 1,000 to 200. ___ %

10) From 144 to 36. ___ %

Calculate each percent of change word problem.

11) Bob got a raise, and his hourly wage increased from $42 to $63. What is the percent increase? ____ %

12) The price of a pair of shoes increases from $50 to $61. What is the percent increase? ___ %

13) At a coffee shop, the price of a cup of coffee increased from $4.80 to $5.76. What is the percent increase in the cost of the coffee? _____ %

14) 51 cm are cut from 85 cm board. What is the percent decrease in length? _____ %

15) In a class, the number of students has been increased from 54 to 81. What is the percent increase? _____ %

16) The price of gasoline rises from $24.40 to $30.50 in one month. By what percent did the gas price rise? _____ %

17) A shirt was originally priced at $38. It went on sale for $24.70. What was the percent that the shirt was discounted? _____ %

Simple Interest

Determine the simple interest for these loans.

1) \$480 at 11% for 3 years. \$ ________
2) \$4,200 at 7% for 4 years. \$ ________
3) \$2,500 at 20% for 3 years. \$ _______
4) \$6,800 at 3.9% for 4 months. \$ ____
5) \$800 at 6% for 7 months. \$ _______
6) \$36,000 at 4.2% for 6 years. \$ _____
7) \$6,500 at 7% for 4 years. \$ _______
8) \$850 at 9.5% for 2 years. \$ ________
9) \$1,200 at 5.8% for 9 months. \$ ____
10) \$3,000 at 4.5% for 7 years. \$ _____

Calculate each simple interest word problem.

11) A new car, valued at \$22,000, depreciates at 8.5% per year. What is the value of the car one year after purchase? \$____________

12) Sara puts \$9,000 into an investment yielding 6% annual simple interest; she left the money in for three years. How much interest does Sara get at the end of those three years? \$____________

13) A bank is offering 12% simple interest on a savings account. If you deposit \$16,400, how much interest will you earn in two years? \$____________

14) \$720 interest is earned on a principal of \$6,000 at a simple interest rate of 4% interest per year. For how many years was the principal invested? _____

15) In how many years will \$2,200 yield an interest of \$440 at 4% simple interest? _____

16) Jim invested \$8,000 in a bond at a yearly rate of 4.5%. He earned \$1,440 in interest. How long was the money invested? _____

Answers of Worksheets

Simplifying Ratios

1) 3: 4
2) 1: 10
3) 4: 7
4) 1: 3
5) 1: 10
6) 1: 8
7) 2: 8
8) 2: 5
9) 1: 6
10) 7: 9
11) 2: 3
12) 7: 2
13) 10: 1
14) 3: 4
15) 1: 8
16) 5: 7
17) 7: 9
18) 2: 1
19) 1: 3
20) 7: 1
21) 1: 2
22) 5: 4
23) 1: 2
24) 1: 30
25) $\frac{1}{2}$
26) $\frac{3}{5}$
27) $\frac{3}{7}$
28) $\frac{1}{3}$
29) $\frac{1}{3}$
30) $\frac{3}{14}$
31) $\frac{1}{2}$
32) $\frac{1}{5}$
33) $\frac{5}{12}$
34) $\frac{2}{9}$
35) $\frac{11}{13}$
36) $\frac{4}{9}$
37) $\frac{1}{12}$
38) $\frac{3}{11}$
39) $\frac{1}{15}$
40) $\frac{1}{4}$
41) $\frac{4}{3}$
42) $\frac{1}{5}$
43) $\frac{22}{41}$
44) $\frac{1}{9}$
45) $\frac{3}{5}$

Proportional Ratios

1) 18
2) 50
3) 9
4) 14
5) 14
6) 56
7) 100
8) 25
9) 70
10) 72
11) 200
12) 77
13) Yes
14) Yes
15) Yes
16) No
17) No
18) No
19) Yes
20) Yes
21) No
22) No
23) Yes
24) Yes
25) 40
26) 256
27) 7
28) 42
29) 63
30) 65
31) 20
32) 104
33) 55
34) 64
35) 126
36) 126

Similarity and ratios

1) 15
2) 5
3) 15
4) 13
5) 150 feet
6) 12.5 meters
7) 50 feet
8) 307.2 miles

Ratio and Rates Word Problems

1) 2: 3
2) 315

3) The ratio for both classes is 7 to 4.	6) $51.75
4) Walmart is a better buy.	7) 56
5) 720, the rate is 30 per hour.	8) 15

Percentage Calculations

1) 1.8	11) 180	21) 8%
2) 6.4	12) 260	22) 32%
3) 2.88	13) 280	23) 2%
4) 5.12	14) 39.15	24) 80%
5) 31	15) 57.6	25) 16%
6) 19.6	16) 33	26) 7%
7) 3	17) 46.2	27) 14%
8) 21	18) 18.8	28) 42%
9) 40	19) 20%	29) 76%
10) 13.8	20) 50%	30) 40

Percent Problems

1) 20%	9) 15%	17) 63%
2) 50%	10) 65%	18) 580%
3) 5%	11) 25%	19) 62.5%
4) 12%	12) 20%	20) 15 ounces
5) 2.5%	13) 10%	21) 80
6) 120%	14) 225%	22) 80%
7) 25%	15) 450%	23) 30%
8) 80%	16) 30%	

Discount, Tax and Tip

1) $453.60	7) $315.00	13) $121.50
2) $291.20	8) $358.80	14) $47.04
3) $861.00	9) $43.50	15) $200.00
4) $16,368.80	10) $631.40	16) $112.50
5) $272.50	11) $51.75	17) $69.30
6) $641,340	12) $38.40	18) $714.00

Percent of Change

1) 200%
2) 400%
3) 500%
4) 200%
5) 200%
6) 80%
7) 200%
8) 50%
9) 80%
10) 75%
11) 50%
12) 22%
13) 20%
14) 60%
15) 50%
16) 25%
17) 35%

Simple Interest

1) $158.40
2) $1,176.00
3) $1,500.00
4) $88.40
5) $28.00
6) $9,072.00
7) $1,820.00
8) $161.50
9) $52.20
10) $945.00
11) $20,130.00
12) $1,620.00
13) $3,936.00
14) 3 years
15) 5 years
16) 4 years

Chapter 4 :

Exponents and Radicals Expressions

Topics that you'll practice in this chapter:

- ✓ Multiplication Property of Exponents
- ✓ Zero and Negative Exponents
- ✓ Division Property of Exponents
- ✓ Powers of Products and Quotients
- ✓ Negative Exponents and Negative Bases
- ✓ Scientific Notation
- ✓ Square Roots
- ✓ Simplifying Radical Expressions
- ✓ Simplifying Radical Expressions Involving Fractions
- ✓ Multiplying Radical Expressions
- ✓ Adding and Subtracting Radical Expressions

Love is anterior to life, posterior to death, initial of creation, and the exponent of breath.

Emily Dickinson

Multiplication Property of Exponents

Simplify and write the answer in exponential form.

1) $4 \times 4^5 =$

2) $8^4 \times 8 =$

3) $7^3 \times 7^3 =$

4) $9^2 \times 9^2 =$

5) $2^2 \times 2^4 \times 2 =$

6) $5 \times 5^3 \times 5^3 =$

7) $4^3 \times 4^2 \times 4 \times 4 =$

8) $5x \times x =$

9) $x^3 \times x^3 =$

10) $x^7 \times x^2 =$

11) $x^4 \times x^3 \times x^2 =$

12) $10x \times 3x =$

13) $4x^3 \times 4x^3 =$

14) $7x^3 \times x =$

15) $3x^2 \times 4x^2 \times x^2 =$

16) $5x^4 \times x^4 =$

17) $2x^8 \times 2x =$

18) $6x \times x^5 =$

19) $4x^2 \times 6x^6 =$

20) $5yx^3 \times 4x =$

21) $7x^3 \times y^5x^7 =$

22) $y^2x^3 \times y^5x^4 =$

23) $3x^5 \times 4x^3y^4 =$

24) $4x^4 \times 9x^2y^5 =$

25) $5x^3y^4 \times 6x^8y^2 =$

26) $8x^3y^6 \times 4xy^3 =$

27) $2xy^5 \times 6x^3y^3 =$

28) $4x^5y^2 \times 4x^2y^8 =$

29) $7x \times 3y^8x^2 \times y^5 =$

30) $x^3 \times 2y^3x^4 \times 2y =$

31) $3yx^4 \times 3y^4x \times 3xy^3 =$

32) $6y^3 \times 2y^2x^4 \times 10yx^5 =$

Zero and Negative Exponents

Evaluate the following expressions.

1) $1^{-5} =$

2) $4^{-1} =$

3) $0^{10} =$

4) $1^{15} =$

5) $5^{-2} =$

6) $3^{-3} =$

7) $9^{-1} =$

8) $10^{-2} =$

9) $12^{-2} =$

10) $2^{-5} =$

11) $3^{-4} =$

12) $2^{-4} =$

13) $6^{-3} =$

14) $10^{-3} =$

15) $30^{-1=}$

16) $15^{-2} =$

17) $4^{-3} =$

18) $2^{-7} =$

19) $5^{-3} =$

20) $4^{-4} =$

21) $3^{-5} =$

22) $10^{-4} =$

23) $2^{-10} =$

24) $8^{-3} =$

25) $20^{-2} =$

26) $14^{-2} =$

27) $9^{-3} =$

28) $100^{-2} =$

29) $5^{-4} =$

30) $4^{-6} =$

31) $(\frac{1}{4})^{-3}$

32) $(\frac{1}{6})^{-2} =$

33) $(\frac{1}{7})^{-2} =$

34) $(\frac{2}{3})^{-3} =$

35) $(\frac{1}{13})^{-2} =$

36) $(\frac{7}{12})^{-2} =$

37) $(\frac{1}{6})^{-3} =$

38) $(\frac{1}{300})^{-2} =$

39) $(\frac{2}{9})^{-2} =$

40) $(\frac{7}{5})^{-1} =$

41) $(\frac{13}{23})^{0} =$

42) $(\frac{1}{4})^{-5} =$

Division Property of Exponents

Simplify.

1) $\frac{5^6}{5^7} =$

2) $\frac{8^8}{8^6} =$

3) $\frac{4^5}{4} =$

4) $\frac{3}{3^5} =$

5) $\frac{x}{x^6} =$

6) $\frac{3\times3^2}{3^2\times3^5} =$

7) $\frac{9^4}{9^2} =$

8) $\frac{10\times10^9}{10^2\times10^7} =$

9) $\frac{7^5\times7^7}{7^4\times7^8} =$

10) $\frac{15x}{30x^6} =$

11) $\frac{3x^9}{4x^4} =$

12) $\frac{15x^8}{10x^9} =$

13) $\frac{42x^5}{6y^9} =$

14) $\frac{36y^8}{4x^4y^5} =$

15) $\frac{2x^7}{9x} =$

16) $\frac{49x^8y^6}{7x^9} =$

17) $\frac{48x^2}{24x^6y^{12}} =$

18) $\frac{30yx^5}{6yx^7} =$

19) $\frac{19x^7y}{38x^{12}y^4} =$

20) $\frac{9x^8}{63x^8} =$

21) $\frac{9x^{-9}}{4x^{-3}} =$

Powers of Products and Quotients

Simplify.

1) $(4^3)^2 =$

2) $(2^3)^4 =$

3) $(2 \times 2^3)^2 =$

4) $(5 \times 5^5)^6 =$

5) $(19^4 \times 19^2)^3 =$

6) $(2^3 \times 2^4)^4 =$

7) $(5 \times 5^2)^2 =$

8) $(4^4)^4 =$

9) $(8x^5)^2 =$

10) $(3x^2y^4)^4 =$

11) $(7x^5y^2)^2 =$

12) $(5x^4y^4)^3 =$

13) $(2x^3y^3)^5 =$

14) $(10x^3y^4)^3 =$

15) $(13y^3y)^2 =$

16) $(5x^6x^4)^2 =$

17) $(6x^7y^6)^3 =$

18) $(12x^5x^7)^2 =$

19) $(2x^4 \times 2x)^4 =$

20) $(2x^4y^3)^5 =$

21) $(15x^7y^2)^2 =$

22) $(8x^3y^5)^3 =$

23) $(3x \times 2y^2)^4 =$

24) $(\frac{4x}{x^5})^2 =$

25) $\left(\frac{x^4y^5}{x^3y^5}\right)^9 =$

26) $\left(\frac{36xy}{6x^5}\right)^3 =$

27) $\left(\frac{x^7}{x^8y^2}\right)^6 =$

28) $\left(\frac{xy^4}{x^3y^6}\right)^{-3} =$

29) $\left(\frac{5xy^8}{x^3}\right)^2 =$

30) $\left(\frac{xy^6}{2xy^3}\right)^{-4} =$

Negative Exponents and Negative Bases

Simplify.

1) $-9^{-1} =$

2) $-9^{-2} =$

3) $-2^{-5} =$

4) $-x^{-7} =$

5) $11x^{-1} =$

6) $-8x^{-3} =$

7) $-12x^{-5} =$

8) $-9x^{-8}y^{-6} =$

9) $32x^{-5}y^{-1} =$

10) $10a^{-9}b^{-3} =$

11) $-17x^{4}y^{-6} =$

12) $-\frac{25}{x^{-5}} =$

13) $-\frac{13x}{a^{-7}} =$

14) $(-\frac{1}{3})^{-4} =$

15) $(-\frac{3}{4})^{-2} =$

16) $-\frac{14}{a^{-6}b^{-3}} =$

17) $-\frac{7x}{x^{-8}} =$

18) $-\frac{a^{-9}}{b^{-5}} =$

19) $-\frac{11}{x^{-5}} =$

20) $\frac{8b}{-16c^{-6}} =$

21) $\frac{12ab}{a^{-4}b^{-3}} =$

22) $-\frac{8n^{-4}}{32p^{-7}} =$

23) $\frac{16ab^{-6}}{-6c^{-5}} =$

24) $(\frac{10a}{5c})^{-4} =$

25) $(-\frac{12x}{4yz})^{-3} =$

26) $\frac{8ab^{-7}}{-5c^{-3}} =$

27) $(-\frac{x^{4}}{x^{5}})^{-5} =$

28) $(-\frac{x^{-2}}{7x^{3}})^{-2} =$

29) $(-\frac{x^{-4}}{x^{2}})^{-6} =$

Scientific Notation

Write each number in scientific notation.

1) 0.223 =

2) 0.09 =

3) 4.5 =

4) 900 =

5) 2,000 =

6) 0.006 =

7) 33 =

8) 9,400 =

9) 1,470 =

10) 52,000 =

11) 8,000,000 =

12) 0.00009 =

13) 2,158,000 =

14) 0.0039 =

15) 0.000075 =

16) 4,300,000 =

17) 130,000 =

18) 4,000,000,000 =

19) 0.00009 =

20) 0.0039 =

Write each number in standard notation.

21) $4 \times 10^{-1} =$

22) $1.2 \times 10^{-3} =$

23) $2.7 \times 10^{5} =$

24) $6 \times 10^{-4} =$

25) $3.6 \times 10^{-3} =$

26) $5.5 \times 10^{5} =$

27) $3.2 \times 10^{4} =$

28) $3.88 \times 10^{6} =$

29) $7 \times 10^{-6} =$

30) $4.2 \times 10^{-7} =$

Square Roots

Find the value each square root.

1) $\sqrt{16}$ = ____
2) $\sqrt{25}$ = ____
3) $\sqrt{1}$ = ____
4) $\sqrt{64}$ = ____
5) $\sqrt{0}$ = ____
6) $\sqrt{196}$ = ____
7) $\sqrt{4}$ = ____
8) $\sqrt{256}$ = ____
9) $\sqrt{36}$ = ____
10) $\sqrt{289}$ = ____
11) $\sqrt{169}$ = ____
12) $\sqrt{144}$ = ____
13) $\sqrt{100}$ = ____
14) $\sqrt{1{,}600}$ = ____
15) $\sqrt{2{,}500}$ = ____
16) $\sqrt{324}$ = ____
17) $\sqrt{529}$ = ____
18) $\sqrt{20}$ = ____
19) $\sqrt{625}$ = ____
20) $\sqrt{18}$ = ____
21) $\sqrt{50}$ = ____
22) $\sqrt{1{,}024}$ = ____
23) $\sqrt{160}$ = ____
24) $\sqrt{32}$ = ____

Evaluate.

25) $\sqrt{4} \times \sqrt{25}$ = ________
26) $\sqrt{36} \times \sqrt{49}$ = ________
27) $\sqrt{6} \times \sqrt{6}$ = ________
28) $\sqrt{13} \times \sqrt{13}$ = ________
29) $2\sqrt{5} \times 3\sqrt{5}$ = ________
30) $\sqrt{12} \times \sqrt{3}$ = ________
31) $\sqrt{13} + \sqrt{13}$ = ________
32) $\sqrt{10} + 2\sqrt{10}$ = ________
33) $12\sqrt{7} - 10\sqrt{7}$ = ________
34) $4\sqrt{10} \times 2\sqrt{10}$ = ________
35) $5\sqrt{3} \times 8\sqrt{3}$ = ________
36) $6\sqrt{3} - \sqrt{12}$ = ________

Simplifying Radical Expressions

Simplify.

1) $\sqrt{13x^2} =$

2) $\sqrt{75x^2} =$

3) $\sqrt[3]{27a} =$

4) $\sqrt{64x^5} =$

5) $\sqrt{216a} =$

6) $\sqrt[3]{63w^3} =$

7) $\sqrt{192x} =$

8) $\sqrt{125v} =$

9) $\sqrt[3]{128x^2} =$

10) $\sqrt{100x^9} =$

11) $\sqrt{16x^4} =$

12) $\sqrt[3]{500a^5} =$

13) $\sqrt{242} =$

14) $\sqrt{392p^3} =$

15) $\sqrt{8m^6} =$

16) $\sqrt{198x^3y^3} =$

17) $\sqrt{121x^5y^5} =$

18) $\sqrt{16a^6b^3} =$

19) $\sqrt{90x^5y^7} =$

20) $\sqrt[3]{64y^2x^6} =$

21) $10\sqrt{16x^4} =$

22) $6\sqrt{81x^2} =$

23) $\sqrt[3]{56x^2y^6} =$

24) $\sqrt[3]{1{,}000x^5y^7} =$

25) $8\sqrt{50a} =$

26) $\sqrt[4]{625x^8y} =$

27) $\sqrt{24x^4y^5r^3} =$

28) $5\sqrt{36x^4y^5z^8} =$

29) $3\sqrt[3]{343x^9y^7} =$

30) $5\sqrt{81a^5b^2c^9} =$

31) $\sqrt[4]{625x^8y^{16}} =$

Multiplying Radical Expressions

Simplify.

1) $\sqrt{5} \times \sqrt{5} =$

2) $\sqrt{5} \times \sqrt{10} =$

3) $\sqrt{3} \times \sqrt{12} =$

4) $\sqrt{49} \times \sqrt{47} =$

5) $\sqrt{7} \times -2\sqrt{28} =$

6) $3\sqrt{15} \times \sqrt{5} =$

7) $4\sqrt{72} \times \sqrt{2} =$

8) $\sqrt{5} \times -\sqrt{49} =$

9) $\sqrt{55} \times \sqrt{11} =$

10) $7\sqrt{42} \times 2\sqrt{216} =$

11) $\sqrt{45}(5 + \sqrt{5}) =$

12) $\sqrt{13x^2} \times \sqrt{13x^3} =$

13) $-2\sqrt{27} \times \sqrt{3} =$

14) $2\sqrt{13x^4} \times \sqrt{13x^4} =$

15) $\sqrt{14x^3} \times \sqrt{7x^2} =$

16) $-8\sqrt{5x} \times \sqrt{7x^5} =$

17) $-2\sqrt{16x^5} \times 4\sqrt{8x^3} =$

18) $-4\sqrt{32}(8 + \sqrt{32}) =$

19) $\sqrt{32x}\,(10 - \sqrt{2x}) =$

20) $\sqrt{2x}(8\sqrt{x^5} + \sqrt{8}) =$

21) $\sqrt{20r}\,(5 + \sqrt{5}) =$

22) $-4\sqrt{7x} \times 3\sqrt{14x^5} =$

23) $-2\sqrt{12x} \times 3\sqrt{2x}$

24) $-\sqrt{7v^3}\,(-3\sqrt{42v}) =$

25) $(\sqrt{11} - 5)(\sqrt{11} + 5) =$

26) $(-3\sqrt{5} + 3)(\sqrt{5} - 4) =$

27) $(4 - 6\sqrt{3})(-6 + \sqrt{3}) =$

28) $(8 - 3\sqrt{5})(7 - \sqrt{5}) =$

29) $(-1 - \sqrt{3x})(4 + \sqrt{3x}) =$

30) $(-5 + 2\sqrt{7r})(-5 + \sqrt{7r}) =$

31) $(-\sqrt{7n} + 1)(-\sqrt{7} - 5) =$

32) $(-3 + \sqrt{3})(5 - 2\sqrt{3x}) =$

Simplifying Radical Expressions Involving Fractions

Simplify.

1) $\frac{\sqrt{5}}{\sqrt{3}} =$

2) $\frac{\sqrt{18}}{\sqrt{45}} =$

3) $\frac{\sqrt{10}}{5\sqrt{2}} =$

4) $\frac{13}{\sqrt{3}} =$

5) $\frac{12\sqrt{5r}}{\sqrt{m^5}} =$

6) $\frac{11\sqrt{2}}{\sqrt{k}} =$

7) $\frac{6\sqrt{20x^3}}{\sqrt{16x}} =$

8) $\frac{\sqrt{14x^3y^4}}{\sqrt{7x^4y^3}} =$

9) $\frac{1}{1-\sqrt{5}} =$

10) $\frac{1-8\sqrt{a}}{\sqrt{11a}} =$

11) $\frac{\sqrt{a}}{\sqrt{a}+\sqrt{b}} =$

12) $\frac{1-\sqrt{5}}{2-\sqrt{6}} =$

13) $\frac{4+\sqrt{7}}{3-\sqrt{8}} =$

14) $\frac{5}{-3-3\sqrt{3}}$

15) $\frac{7}{2-\sqrt{5}} =$

16) $\frac{\sqrt{7}-\sqrt{3}}{\sqrt{3}-\sqrt{7}} =$

17) $\frac{\sqrt{5}+\sqrt{7}}{\sqrt{7}-\sqrt{5}} =$

18) $\frac{2\sqrt{2}-\sqrt{3}}{3\sqrt{2}+\sqrt{5}} =$

19) $\frac{\sqrt{11}+5\sqrt{3}}{4-\sqrt{11}} =$

20) $\frac{\sqrt{5}+\sqrt{3}}{2-\sqrt{3}} =$

21) $\frac{\sqrt{32a^7b^4}}{\sqrt{2ab^3}} =$

22) $\frac{10\sqrt{21x^5}}{5\sqrt{x^3}} =$

Adding and Subtracting Radical Expressions

Simplify.

1) $\sqrt{2}+\sqrt{8}=$

2) $3\sqrt{50}+4\sqrt{2}=$

3) $2\sqrt{12}-4\sqrt{3}=$

4) $5\sqrt{32}-5\sqrt{2}=$

5) $3\sqrt{75}-5\sqrt{3}=$

6) $-\sqrt{72}-4\sqrt{2}=$

7) $-7\sqrt{16}-4\sqrt{25}=$

8) $8\sqrt{24}+2\sqrt{6}=$

9) $10\sqrt{49}-7\sqrt{100}=$

10) $-7\sqrt{5}+9\sqrt{45}=$

11) $-15\sqrt{12}+14\sqrt{48}=$

12) $20\sqrt{4}-2\sqrt{25}=$

13) $-2\sqrt{20}+7\sqrt{5}=$

14) $8\sqrt{7}-2\sqrt{63}=$

15) $5\sqrt{44}+3\sqrt{11}=$

16) $3\sqrt{27}-5\sqrt{48}=$

17) $\sqrt{144}-\sqrt{81}=$

18) $3\sqrt{20}-6\sqrt{5}=$

19) $-2\sqrt{7}+8\sqrt{28}=$

20) $3\sqrt{75}-2\sqrt{3}=$

21) $5\sqrt{27}-3\sqrt{3}=$

22) $-7\sqrt{30}+6\sqrt{120}=$

23) $-7\sqrt{24}-2\sqrt{6}=$

24) $-\sqrt{32x}+4\sqrt{2x}=$

25) $\sqrt{7y^2}+y\sqrt{112}=$

26) $\sqrt{45mn^2}+2n\sqrt{5m}=$

27) $-4\sqrt{12a}-4\sqrt{3a}=$

28) $-5\sqrt{15ab}-2\sqrt{60ab}=$

29) $\sqrt{45x^2y}+x\sqrt{20y}=$

30) $2\sqrt{7\text{a}}+4\sqrt{63\text{a}}=$

Answers of Worksheets

Multiplication Property of Exponents

1) 4^6
2) 8^5
3) 7^6
4) 9^4
5) 2^7
6) 5^7
7) 4^7
8) $5x^2$
9) x^6
10) x^9
11) x^9
12) $30x^2$
13) $16x^6$
14) $7x^4$
15) $12x^6$
16) $5x^8$
17) $4x^9$
18) $6x^6$
19) $24x^8$
20) $20x^4y$
21) $7x^{10}y^5$
22) x^7y^7
23) $12x^8y^4$
24) $36x^6y^5$
25) $30x^{11}y^6$
26) $32x^4y^9$
27) $12x^4y^8$
28) $16x^7y^{10}$
29) $21x^3y^{13}$
30) $4x^7y^4$
31) $27x^6y^8$
32) $120x^9y^6$

Zero and Negative Exponents

1) 1
2) $\frac{1}{4}$
3) 0
4) 1
5) $\frac{1}{25}$
6) $\frac{1}{27}$
7) $\frac{1}{9}$
8) $\frac{1}{100}$
9) $\frac{1}{144}$
10) $\frac{1}{32}$
11) $\frac{1}{81}$
12) $\frac{1}{16}$
13) $\frac{1}{216}$
14) $\frac{1}{1,000}$
15) $\frac{1}{30}$
16) $\frac{1}{225}$
17) $\frac{1}{64}$
18) $\frac{1}{128}$
19) $\frac{1}{125}$
20) $\frac{1}{256}$
21) $\frac{1}{243}$
22) $\frac{1}{10,000}$
23) $\frac{1}{1,024}$
24) $\frac{1}{512}$
25) $\frac{1}{400}$
26) $\frac{1}{196}$
27) $\frac{1}{729}$
28) $\frac{1}{10,000}$
29) $\frac{1}{625}$
30) $\frac{1}{4,096}$
31) 64
32) 36
33) 49
34) $\frac{27}{8}$
35) 169
36) $\frac{144}{49}$
37) 216
38) 90,000
39) $\frac{81}{4}$
40) $\frac{5}{7}$
41) 1
42) 1,024

Division Property of Exponents

1) $\frac{1}{5}$
2) 8^2
3) 4^4
4) $\frac{1}{3^4}$
5) $\frac{1}{x^5}$
6) $\frac{1}{3^4}$
7) 9^2
8) 10
9) 1
10) $\frac{1}{2x^5}$
11) $\frac{3x^5}{4}$
12) $\frac{3}{2x}$
13) $\frac{7x^5}{y^9}$
14) $\frac{9y^3}{x^4}$
15) $\frac{2x^6}{9}$

16) $\frac{7y^6}{x}$
17) $\frac{2}{x^4y^{12}}$
18) $\frac{5}{x^2}$
19) $\frac{1}{2x^5y^3}$
20) $\frac{1}{7}$
21) $\frac{9}{4x^6}$

Powers of Products and Quotients

1) 4^6
2) 2^{12}
3) 2^8
4) 5^{36}
5) 19^{18}
6) 2^{28}
7) 5^6
8) 4^{16}
9) $64x^{10}$
10) $81x^8y^{16}$
11) $49x^{10}y^4$
12) $125x^{12}y^{12}$
13) $32x^{15}y^{15}$
14) $1{,}000x^9y^{12}$
15) $169y^8$
16) $25x^{20}$
17) $216x^{21}y^{18}$
18) $144x^{24}$
19) $256x^{20}$
20) $32x^{20}y^{15}$
21) $225x^{14}y^4$
22) $512x^9y^{15}$
23) $1{,}296x^4y^8$
24) $\frac{16}{x^8}$
25) x^9
26) $\frac{216y^3}{x^{12}}$
27) $\frac{1}{x^6y^{12}}$
28) x^6y^6
29) $\frac{25y^{16}}{x^4}$
30) $\frac{16}{y^{12}}$

Negative Exponents and Negative Bases

1) $-\frac{1}{9}$
2) $-\frac{1}{81}$
3) $-\frac{1}{32}$
4) $-\frac{1}{x^7}$
5) $\frac{11}{x}$
6) $-\frac{8}{x^3}$
7) $-\frac{12}{x^5}$
8) $-\frac{9}{x^8y^6}$
9) $\frac{32}{x^5y}$
10) $\frac{10}{a^9b^3}$
11) $-\frac{17x^4}{y^6}$
12) $-25x^5$
13) $-13xa^7$
14) 81
15) $\frac{16}{9}$
16) $-14a^6b^3$
17) $-7x^9$
18) $-\frac{b^5}{a^9}$
19) $-11x^5$
20) $-\frac{bc^6}{2}$
21) $12a^5b^4$
22) $-\frac{p^7}{4n^4}$
23) $-\frac{8ac^5}{3b^6}$
24) $\frac{c^4}{16a^4}$
25) $\frac{y^3z^3}{27x^3}$
26) $-\frac{8ac^3}{5b^7}$
27) $-x^5$
28) $49x^{10}$
29) x^{36}

Scientific Notation

1) 2.23×10^{-1}
2) 9×10^{-2}
3) 4.5×10^{0}
4) 9×10^{2}
5) 2×10^{3}
6) 6×10^{-3}
7) 3.3×10^{1}
8) 9.4×10^{3}
9) 1.47×10^{3}
10) 5.2×10^{4}
11) 8×10^{6}
12) 9×10^{-5}
13) 2.158×10^{6}
14) 3.9×10^{-3}
15) 7.5×10^{-5}
16) 4.3×10^{6}
17) 1.3×10^{5}
18) 4×10^{9}
19) 9×10^{-5}
20) 3.9×10^{-3}
21) 0.4
22) 0.0012
23) 270,000
24) 0.0006
25) 0.0036
26) 550,000
27) 32,000
28) 3,880,000
29) 0.000007
30) 0.00000042

Square Roots

1) 4
2) 5
3) 1
4) 8
5) 0
6) 14
7) 2
8) 16
9) 6
10) 17
11) 13
12) 12
13) 10
14) 40
15) 50
16) 18
17) 23
18) $2\sqrt{5}$
19) 25
20) $3\sqrt{2}$
21) $5\sqrt{2}$
22) 32
23) $4\sqrt{10}$
24) $4\sqrt{2}$
25) 10
26) 42
27) 6
28) 13
29) 30
30) 6
31) $2\sqrt{13}$
32) $3\sqrt{10}$
33) $2\sqrt{7}$
34) 80
35) 120
36) $4\sqrt{3}$

Simplifying radical expressions

1) $x\sqrt{13}$
2) $5x\sqrt{3}$
3) $3\sqrt[3]{a}$
4) $8x^2\sqrt{x}$
5) $6\sqrt{6a}$
6) $w\sqrt[3]{63}$
7) $8\sqrt{3x}$
8) $5\sqrt{5v}$
9) $4\sqrt[3]{2x^2}$
10) $10x^4\sqrt{x}$
11) $4x^2$
12) $5a\sqrt[3]{4a^2}$
13) $11\sqrt{2}$
14) $14p\sqrt{2p}$
15) $2m^3\sqrt{2}$
16) $3x.y\sqrt{22xy}$
17) $11x^2y^2\sqrt{xy}$
18) $4a^3b\sqrt{b}$
19) $3x^2y^3\sqrt{10xy}$
20) $4x^2\sqrt[3]{y^2}$
21) $40x^2$
22) $54x$
23) $2y^2\sqrt[3]{7x^2}$
24) $10xy^2\sqrt[3]{x^2y}$

25) $40\sqrt{2a}$

26) $5x^2\sqrt[4]{y}$

27) $2x^2y^2r\sqrt{6yr}$

28) $30x^2y^2z^4\sqrt{y}$

29) $21x^3y^2\sqrt[3]{y}$

30) $45a^2bc^4\sqrt{ac}$

31) $5x^2y^4$

Multiplying radical expressions

1) 5
2) $5\sqrt{2}$
3) 6
4) $7\sqrt{47}$
5) -28
6) $15\sqrt{3}$
7) 48
8) $-5\sqrt{7}$
9) $11\sqrt{5}$
10) $504\sqrt{7}$
11) $15\sqrt{5}+15$
12) $13x^2\sqrt{x}$
13) -18
14) $26x^4$
15) $7x^2\sqrt{2x}$
16) $-8x^3\sqrt{35}$
17) $-64x^4\sqrt{2}$
18) $-128\sqrt{2}-128$
19) $40\sqrt{2x}-8x$
20) $8x^3\sqrt{2}+4\sqrt{x}$
21) $10\sqrt{5r}+10\sqrt{r}$
22) $-84x^3\sqrt{2}$
23) $-12\sqrt{6}x$
24) $21v^2\sqrt{6}$
25) -14
26) $15\sqrt{5}-27$
27) $40\sqrt{3}-42$
28) $71-29\sqrt{5}$
29) $-3x-5\sqrt{3x}-4$
30) $14r-15\sqrt{7r}+25$
31) $7\sqrt{n}+5\sqrt{7n}-\sqrt{7}-5$
32) $-15+6\sqrt{3x}+5\sqrt{3}-6\sqrt{x}$

Simplifying radical expressions involving fractions

1) $\frac{\sqrt{15}}{3}$
2) $\frac{9\sqrt{10}}{45}=\frac{\sqrt{10}}{5}$
3) $\frac{\sqrt{20}}{10}=\frac{\sqrt{5}}{5}$
4) $\frac{13\sqrt{3}}{3}$
5) $\frac{12\sqrt{5mr}}{m^3}$
6) $\frac{11\sqrt{2k}}{k}$
7) $3x\sqrt{5}$
8) $\frac{\sqrt{2x}}{xy}$
9) $\frac{-1-\sqrt{5}}{4}$
10) $\frac{\sqrt{11a}-8a\sqrt{11}}{11a}$

11) $\frac{a-\sqrt{ab}}{a-b}$

12) $\frac{\sqrt{30}+2\sqrt{5}-\sqrt{6}-2}{2}$

13) $12+8\sqrt{2}+3\sqrt{7}+2\sqrt{14}$

14) $-\frac{5(\sqrt{3}-1)}{6}$

15) $-14-7\sqrt{5}$

16) -1

17) $6+\sqrt{35}$

18) $\frac{12-2\sqrt{10}-3\sqrt{6}+\sqrt{15}}{13}$

19) $\frac{4\sqrt{11}+11+20\sqrt{3}+5\sqrt{33}}{5}$

20) $2\sqrt{5}+3+\sqrt{15}+2\sqrt{3}$

21) $4a^3\sqrt{b}$

22) $2x\sqrt{21}$

Adding and subtracting radical expressions

1) $3\sqrt{2}$
2) $19\sqrt{2}$
3) 0
4) $15\sqrt{2}$
5) $10\sqrt{3}$
6) $-10\sqrt{2}$
7) -48
8) $18\sqrt{6}$
9) 0
10) $20\sqrt{5}$
11) $26\sqrt{3}$
12) 30
13) $3\sqrt{5}$
14) $2\sqrt{7}$
15) $13\sqrt{11}$
16) $-11\sqrt{3}$
17) 3
18) 0
19) $14\sqrt{7}$
20) $13\sqrt{3}$
21) $12\sqrt{3}$
22) $5\sqrt{30}$
23) $-16\sqrt{6}$
24) 0
25) $5y\sqrt{7}$
26) $5n\sqrt{5m}$
27) $-12\sqrt{3a}$
28) $-9\sqrt{15ab}$
29) $5x\sqrt{5y}$
30) $14\sqrt{7a}$

Chapter 5 :

Algebraic Expressions

Topics that you'll practice in this chapter:

- ✓ Simplifying Variable Expressions
- ✓ Simplifying Polynomial Expressions
- ✓ Translate Phrases into an Algebraic Statement
- ✓ The Distributive Property
- ✓ Evaluating One Variable Expressions
- ✓ Evaluating Two Variables Expressions
- ✓ Combining like Terms

I want freedom for the full expression of my personality.

Mahatma Gandhi

Simplifying Variable Expressions

Simplify each expression.

1) $3(x+5) =$

2) $(-4)(7x-5) =$

3) $11x+5-6x =$

4) $-4-2x^2-6x^2 =$

5) $7+13x^2+3 =$

6) $3x^2+7x+15x^2 =$

7) $3x^2-12x^2+4x =$

8) $4x^2-8x-2x =$

9) $6x+7(3-4x) =$

10) $8x+4(15x-3) =$

11) $6(-3x-9)-17 =$

12) $-11x^2-(-5x) =$

13) $2x+7+5-8x =$

14) $7+6x-11-5x =$

15) $27x+8-13-5x =$

16) $(-11)(-5x+2)-41x =$

17) $19x-4(4-2x) =$

18) $16x+3(3x+6)+10 =$

19) $5(-2x-4)-13x =$

20) $16x-3x(x+10) =$

21) $17x+5x(2-4x) =$

22) $5x(-4x-7)+20x =$

23) $25x-19+4x^2 =$

24) $6x(x-11)+25 =$

25) $4x-5+15x+3x^2 =$

26) $-7x^2-11x-9x =$

27) $10x-9x^2-3x^2-7 =$

28) $13+3x^2-9x^2-21x =$

29) $22x+10x^2-15x+17 =$

30) $4x^2+25x+21x^2 =$

31) $29-12x^2-23x-4x^2 =$

32) $22x-19x-9x^2+30 =$

Simplifying Polynomial Expressions

Simplify each polynomial.

1) $(2x^3 + 8x^2) - (11x + 3x^2) =$ ______________

2) $(2x^5 + 7x^3) - (5x^3 + 11x^2) =$ ______________

3) $(41x^4 + 5x^2) - (4x^2 + 20x^4) =$ ______________

4) $13x - 8x^2 + 4(4x^2 + 3x^3) =$ ______________

5) $(4x^3 - 22) + 5(3x^2 - 6x^3) =$ ______________

6) $(4x^3 - 3x) - 5(2x^3 + x^4) =$ ______________

7) $5(5x - 2x^3) - 2(8x^3 + 5x^2) =$ ______________

8) $(3x^2 - 10x) - (5x^3 + 14x^2) =$ ______________

9) $5x^3 - (3x^4 + 5x) + 2x^2 =$ ______________

10) $11x^4 - (3x^2 + 5x) + 7x =$ ______________

11) $(6x^2 - 3x^4) - (10x^4 + 3x^2) =$ ______________

12) $2x^2 - 7x^3 + 19x^4 - 22x^3 =$ ______________

13) $10x^2 - x^4 + 4x^4 - 32x^3 =$ ______________

14) $-5x^2 + 17x^3 - 8x^2 - 6x =$ ______________

15) $x^4 - 11x^5 - 30x^4 + 5x^2 =$ ______________

16) $21x^3 + 13x - 5x^2 - 11x^3 =$ ______________

Translate Phrases into an Algebraic Statement

Write an algebraic expression for each phrase.

1) 9 multiplied by x. ________________

2) Subtract 11 from y. ________________

3) 19 divided by x. ________________

4) 38 decreased by y. ________________

5) Add y to 40. ________________

6) The square of 6. ________________

7) x raised to the fifth power. ________________

8) The sum of six and a number. ________________

9) The difference between fifty–seven and y. ________________

10) The quotient of nine and a number. ________________

11) The quotient of the square of x and 25. ________________

12) The difference between x and 6 is 19. ________________

13) 10 times a reduced by the square of b. ________________

14) Subtract the product of a and b from 41. ________________

The Distributive Property

Use the distributive property to simply each expression.

1) $4(1+2x)=$

2) $2(4+7x)=$

3) $3(4x-4)=$

4) $(2x-5)(-6)=$

5) $(-3)(x+6)=$

6) $(4+3x)2=$

7) $(-5)(8-3x)=$

8) $-(-5-7x)=$

9) $(-6x+3)(-3)=$

10) $(-4)(x-7)=$

11) $-(5-3x)=$

12) $3(9+4x)=$

13) $6(4+3x)=$

14) $(-5x+3)2=$

15) $(5-8x)(-3)=$

16) $(-12)(3x+3)=$

17) $(5-3x)6=$

18) $4(2+6x)=$

19) $8(7x-3)=$

20) $(-2x+3)4=$

21) $(7-5x)(-9)=$

22) $(-10)(x-8)=$

23) $(11-4x)3=$

24) $(-6)(10x-4)=$

25) $(3-9x)(-7)=$

26) $(-9)(x+9)=$

27) $(-3+5x)(-7)=$

28) $(-5)(8-10x)=$

29) $12(4x-8)=$

30) $(-10x+13)(-3)=$

31) $(-8)(3x-2)+4(x+5)=$

32) $(-8)(x+4)-(6+5x)=$

Evaluating One Variable Expressions

Evaluate each expression using the value given.

1) $8 - x, x = 5$

2) $x - 9, x = 5$

3) $5x + 4, x = 3$

4) $x - 13, x = -4$

5) $12 - x, x = 4$

6) $x + 2, x = 6$

7) $4x + 8, x = 3$

8) $x + (-7), x = -8$

9) $4x + 5, x = 2$

10) $3x + 9, x = -2$

11) $15 + 3x - 7, x = 2$

12) $17 - 3x, x = 3$

13) $8x - 9, x = 4$

14) $5x + 4, x = -3$

15) $10x + 5, x = 3$

16) $14 - 4x, x = -6$

17) $3(5x + 3), x = 9$

18) $4(-3x - 6), x = 3$

19) $7x - 2x + 12, x = 4$

20) $(5x + 6) \div 2, x = 8$

21) $(x + 18) \div 10, x = 12$

22) $5x - 12 + 3x, x = -3$

23) $(6 - 4x)(-3), x = -4$

24) $9x^2 + 3x - 6, x = 2$

25) $x^2 - 10x, x = -5$

26) $3x(7 - 2x), x = 2$

27) $12x + 6 - 2x^2, x = -4$

28) $(-3)(4x - 8 + 3x), x = 3$

29) $(-6) + \frac{x}{4} + 3x, x = 16$

30) $(-6) + \frac{x}{5}, x = 35$

31) $\left(-\frac{45}{x}\right) - 7 + 2x, x = 9$

32) $\left(-\frac{21}{x}\right) - 12 + 4x, x = 7$

Evaluating Two Variables Expressions

Evaluate each expression using the values given.

1) $2x - 4y$,

$x = 4, y = 1$

2) $3x + 5y$,

$x = -2, y = 2$

3) $-7a + 4b$,

$a = 2, b = 4$

4) $3x + 5 - y$,

$x = 5, y = 6$

5) $3z + 12 - 2k$,

$z = 5, k = 6$

6) $6(-x - 3y)$,

$x = 5, y = -2$

7) $5a + 3b$,

$a = 3, b = 4$

8) $7x \div 3y$,

$x = 3, y = 7$

9) $2x + 15 + 5y$,

$x = -3, y = 1$

10) $5a - (18 - b)$,

$a = 2, b = 8$

11) $2z + 20 + 5k$,

$z = -6, k = 5$

12) $xy + 10 + 4x$,

$x = 3, y = 5$

13) $2x + 4y - 8 + 5$,

$x = 5, y = 2$

14) $\left(-\frac{24}{x}\right) + 3 + 2y$,

$x = 4, y = 6$

15) $(-3)(-3a - 3b)$,

$a = 4, b = 5$

16) $12 + 4x - 7 - y$,

$x = 3, y = 5$

17) $11x + 5 - 8y + 6$,

$x = 5, y = 2$

18) $10 + 2(-4x - 5y)$,

$x = 5, y = 4$

19) $5x + 13 + 6y$,

$x = 5, y = 6$

20) $10a - (7a + 3b) - 11$,

$a = 3, b = 8$

Combining like Terms

Simplify each expression.

1) $11x + 3x + 6 =$

2) $8(2x - 6) =$

3) $18x - 7x + 11 =$

4) $(-4)(6x - 7) =$

5) $22x - 10x - 5 =$

6) $32x - 13 + 8x =$

7) $15 - (8x - 11) =$

8) $-24x + 17 - 11x =$

9) $12x - 8 - 6x + 9 =$

10) $21x + 5 - 36 + 12x =$

11) $28x + 3x - 11 =$

12) $(-3x + 4)5 =$

13) $2 + 4x + 9x - 8 =$

14) $6(2x - 5x) - 4 =$

15) $4(5x + 11) + 3x =$

16) $x - 14 - 11x =$

17) $5(10 + 9x) - 8x =$

18) $42x + 17 - 23x =$

19) $(-7x) + 19 + 20x =$

20) $(-7x) - 33 + 29x =$

21) $4(5x + 3) - 19x =$

22) $5(6 - 2x) - 15x =$

23) $-24x + (11 - 18x) =$

24) $(-9) - (6)(7x + 3) =$

25) $(-1)(8x - 10) - 21x =$

26) $-36x + 14 + 27x - 5x =$

27) $3(-13x + 6) - 17x =$

28) $-5x - 42 + 32x =$

29) $37x - 19x + 15 - 9x =$

30) $3(5x + 7x) - 31 =$

31) $14 - 6x - 15 - 9x =$

32) $-2(-5x - 7x) + 27x =$

Answers of Worksheets

Simplifying Variable Expressions

1) $3x+15$
2) $-28x+20$
3) $5x+5$
4) $-8x^2-4$
5) $13x^2+10$
6) $18x^2+7x$
7) $-9x^2+4x$
8) $4x^2-10x$
9) $-22x+21$
10) $68x-12$
11) $-18x-71$
12) $-11x^2+5x$
13) $-6x+12$
14) $x-4$
15) $22x-5$
16) $14x-22$
17) $27x-16$
18) $25x+28$
19) $-23x-20$
20) $-3x^2-14x$
21) $-20x^2+27x$
22) $-20x^2-15x$
23) $4x^2+25x-19$
24) $6x^2-66x+25$
25) $3x^2+19x-5$
26) $-7x^2-20x$
27) $-12x^2+10x-7$
28) $-6x^2-21x+13$
29) $10x^2+7x+17$
30) $25x^2+25x$
31) $-16x^2-23x+29$
32) $-9x^2+3x+30$

Simplifying Polynomial Expressions

1) $2x^3+5x^2-11x$
2) $2x^5+2x^3-11x^2$
3) $21x^4+x^2$
4) $12x^3+8x^2+13x$
5) $-26x^3+15x^2-22$
6) $-5x^4-6x^3-3x$
7) $-26x^3-10x^2+25x$
8) $-5x^3-11x^2-10x$
9) $-3x^4+5x^3+2x^2-5x$
10) $11x^4-3x^2+2x$
11) $-13x^4+3x^2$
12) $19x^4-29x^3+2x^2$
13) $3x^4-32x^3+10x^2$
14) $17x^3-13x^2-6x$
15) $-11x^5-29x^4+5x^2$
16) $10x^3-5x^2+13x$

Translate Phrases into an Algebraic Statement

1) $9x$
2) $y-11$
3) $\frac{19}{x}$
4) $38-y$
5) $y+40$
6) 6^2
7) x^5
8) $6+x$
9) $57-y$
10) $\frac{9}{x}$
11) $\frac{x^2}{25}$
12) $x-6=19$
13) $10a-b^2$
14) $41-ab$

The Distributive Property

1) $8x+4$
2) $14x+8$
3) $12x-12$
4) $-12x+30$
5) $-3x-18$
6) $6x+8$
7) $15x-40$
8) $7x+5$
9) $18x-9$
10) $-4x+28$
11) $3x-5$
12) $12x+27$

13) $18x + 24$	18) $24x + 8$	23) $-12x + 33$	28) $50x - 40$
14) $-10x + 6$	19) $56x - 24$	24) $-60x + 24$	29) $48x - 96$
15) $24x - 15$	20) $-8x + 12$	25) $63x - 21$	30) $30x - 39$
16) $-36x - 36$	21) $45x - 63$	26) $-9x - 81$	31) $-20x + 36$
17) $-18x + 30$	22) $-10x + 80$	27) $-35x + 21$	32) $-13x - 38$

Evaluating One Variables

1) 3	9) 13	17) 144	25) 75
2) -4	10) 3	18) -60	26) 18
3) 19	11) 14	19) 32	27) -74
4) -17	12) 8	20) 23	28) -39
5) 8	13) 23	21) 3	29) 46
6) 8	14) -11	22) -36	30) 1
7) 20	15) 35	23) -66	31) 6
8) -15	16) 38	24) 36	32) 13

Evaluating Two Variables

1) 4	6) 6	11) 33	16) 12
2) 4	7) 27	12) 37	17) 50
3) 2	8) 1	13) 15	18) -70
4) 14	9) 14	14) 9	19) 74
5) 15	10) 0	15) 81	20) -26

Combining like Terms

1) $14x + 6$	9) $6x + 1$	17) $37x + 50$	25) $-29x + 10$
2) $16x - 48$	10) $33x - 31$	18) $19x + 17$	26) $-14x + 14$
3) $11x + 11$	11) $31x - 11$	19) $13x + 19$	27) $-56x + 18$
4) $-24x + 28$	12) $-15x + 20$	20) $22x - 33$	28) $27x - 42$
5) $12x - 5$	13) $13x - 6$	21) $x + 12$	29) $9x + 15$
6) $40x - 13$	14) $-18x - 4$	22) $-25x + 30$	30) $36x - 31$
7) $-8x + 26$	15) $23x + 44$	23) $-42x + 11$	31) $-15x - 1$
8) $-35x + 17$	16) $-10x - 14$	24) $-42x - 27$	32) $51x$

Chapter 6 :

Equations and Inequalities

Topics that you'll practice in this chapter:

- ✓ One–Step Equations
- ✓ Multi–Step Equations
- ✓ Graphing Single–Variable Inequalities
- ✓ One–Step Inequalities
- ✓ Multi-Step Inequalities
- ✓ Systems of Equations
- ✓ Systems of Equations Word Problems

"Life is a math equation. In order to gain the most, you have to know how to convert negatives into positives." – Anonymous

One–Step Equations

Find the answer for each equation.

1) $3x = 90, x =$ ____

2) $5x = 35, x =$ ____

3) $6x = 24, x =$ ____

4) $24x = 144, x =$ ____

5) $x + 15 = 20, x =$ ____

6) $x - 7 = 4, x =$ ____

7) $x - 9 = 2, x =$ ____

8) $x + 15 = 23, x =$ ____

9) $x - 4 = 13, x =$ ____

10) $12 = 16 + x, x =$ ____

11) $x - 10 = 2, x =$ ____

12) $5 - x = -11, x =$ ____

13) $28 = -6 + x, x =$ ____

14) $x - 20 = -35, x =$ ____

15) $x + 14 = -4, x =$ ____

16) $14 = 28 - x, x =$ ____

17) $7 + x = -7, x =$ ____

18) $x - 16 = 4, x =$ ____

19) $30 = x - 15, x =$ ____

20) $x - 5 = -18, x =$ ____

21) $x - 10 = 24, x =$ ____

22) $x - 20 = -25, x =$ ____

23) $x - 17 = 30, x =$ ____

24) $-70 = x - 28, x =$ ____

25) $x - 9 = 13, x =$ ____

26) $36 = 4x, x =$ ____

27) $x - 35 = 25, x =$ ____

28) $x - 25 = 10, x =$ ____

29) $70 - x = 16, x =$ ____

30) $x - 10 = 14, x =$ ____

31) $17 - x = -13, x =$ __

32) $\mathrm{x} - 9 = -30, \mathrm{x} =$ ____

Multi–Step Equations

Find the answer for each equation.

1) $3x + 3 = 9$

2) $-x + 5 = 12$

3) $4x - 8 = 8$

4) $-(3 - x) = 5$

5) $4x - 8 = 16$

6) $12x - 15 = 9$

7) $2x - 18 = 2$

8) $4x + 8 = 16$

9) $24x + 27 = 75$

10) $-14(3 + x) = 14$

11) $-3(2 + x) = 6$

12) $12 = -(x - 7)$

13) $3(3 - x) = 30$

14) $-15 = -(3x + 6)$

15) $40(3 + x) = 40$

16) $5(x - 10) = 25$

17) $-18 = x + 8x$

18) $3x + 25 = -2x - 10$

19) $7(6 + 3x) = -63$

20) $18 - 3x = -4 - 5x$

21) $4 - 6x = 36 + 2x$

22) $15 + 15x = -5 + 5x$

23) $42 = (-6x) - 7 + 7$

24) $21 = 3x - 21 + 4x$

25) $-18 = -6x - 9 + 3x$

26) $5x - 15 = -29 + 6x$

27) $7x - 18 = 4x + 3$

28) $-7 - 4x = 5(4 - x)$

29) $x - 5 = -5(-3 - x)$

30) $13x - 68 = 15x - 102$

31) $-5x - 3 = -3(9 + 3x)$

32) $-2x - 15 = 6x + 17$

Graphing Single–Variable Inequalities

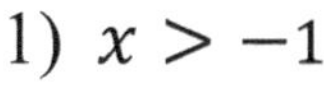 **Draw a graph for each inequality.**

1) $x > -1$

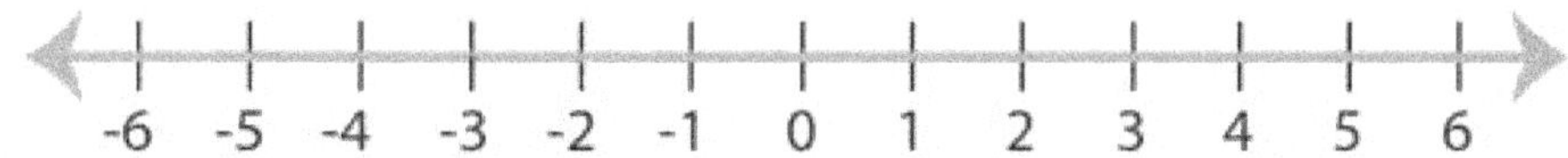

2) $x \leq 2$

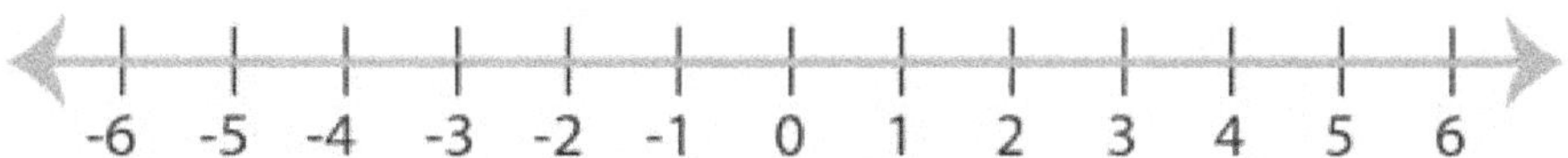

3) $x \geq 0$

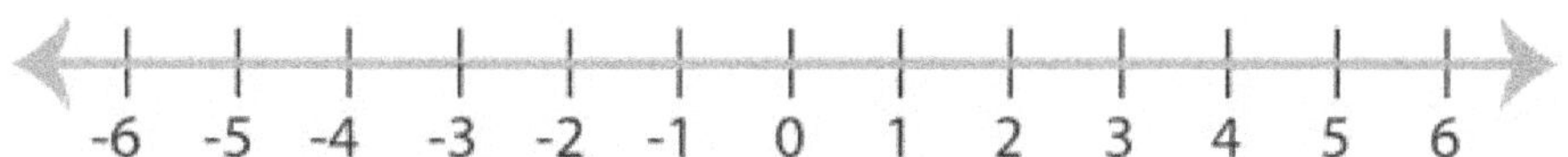

4) $x < -3$

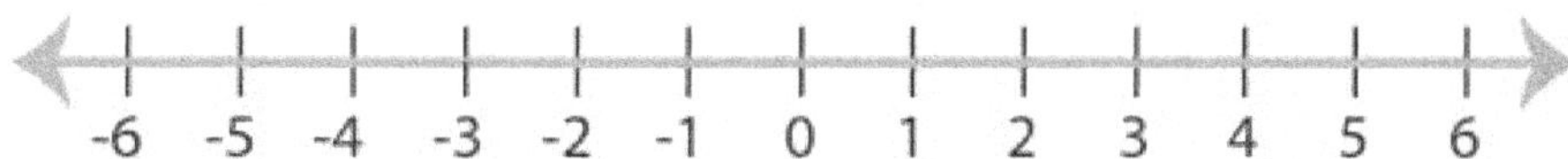

5) $x < \frac{1}{2}$

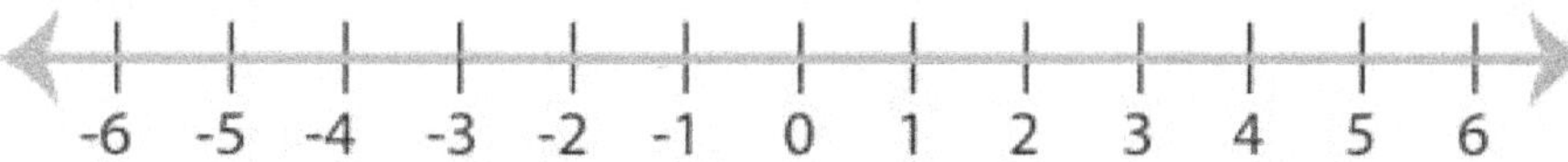

6) $x \leq -2$

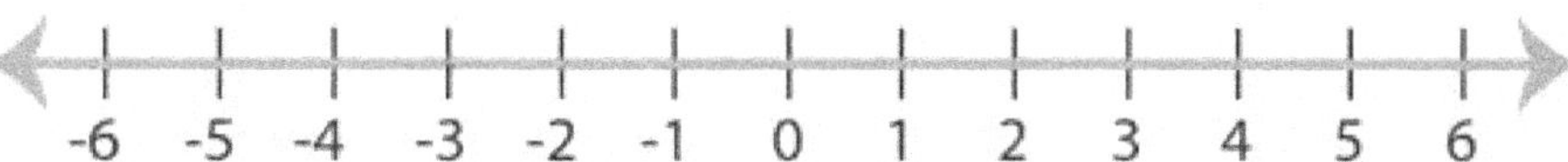

7) $x \leq 3$

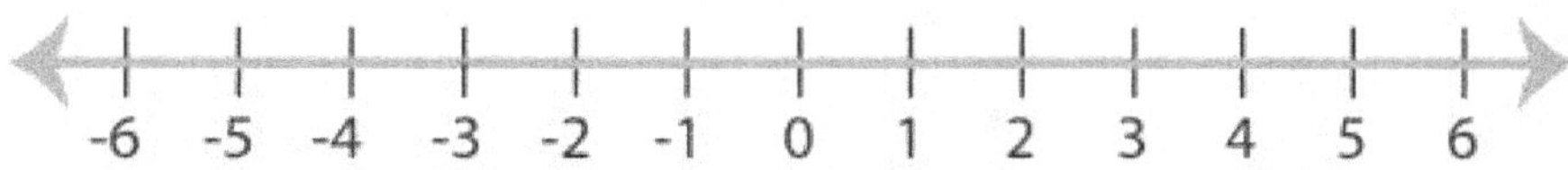

8) $x \geq -\frac{7}{2}$

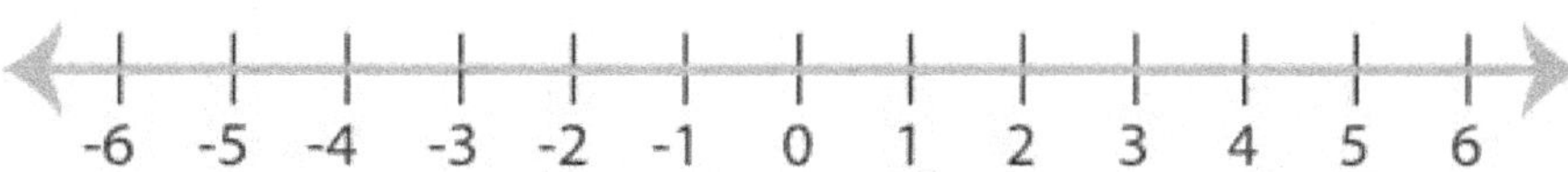

One–Step Inequalities

Find the answer for each inequality and graph it.

1) $x + 4 \geq 4$

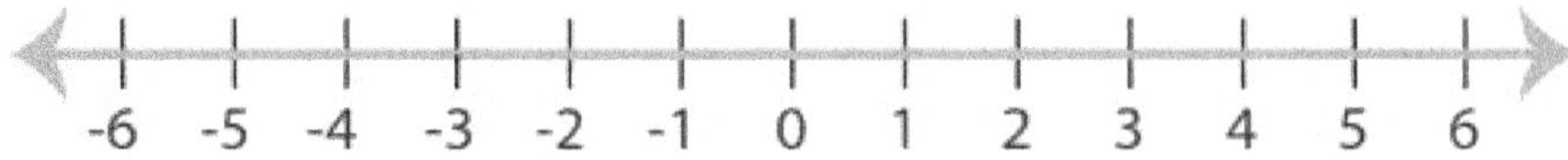

2) $x - 5 \leq 2$

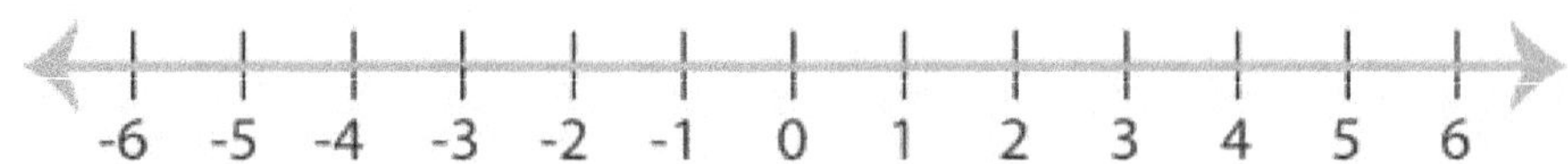

3) $5x > 35$

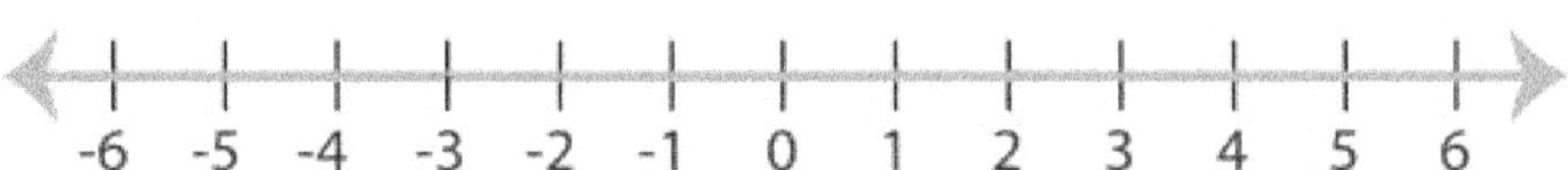

4) $9 + x \leq 11$

5) $x - 5 < -9$

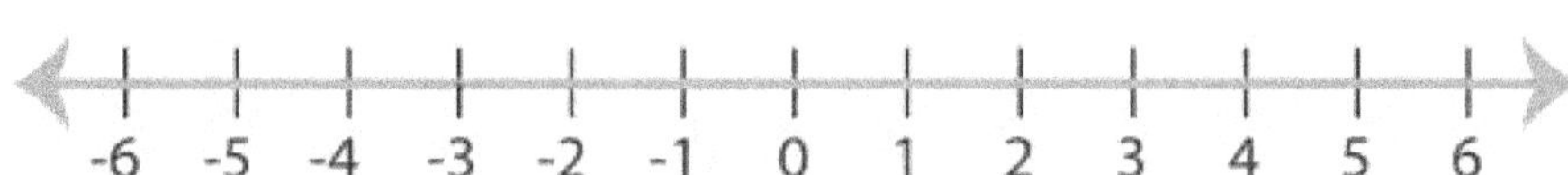

6) $9x \geq 72$

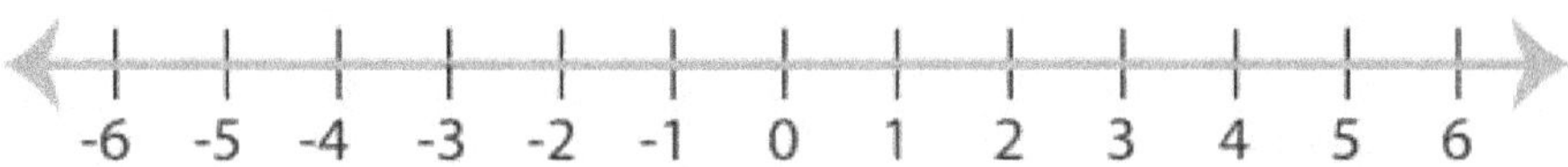

7) $9x \leq 27$

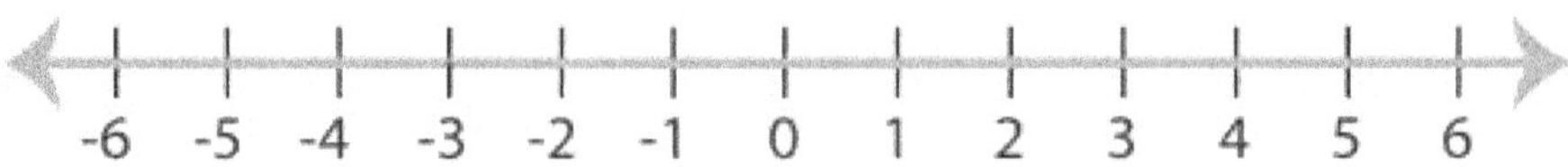

8) $x + 19 > 16$

Multi-Step Inequalities

Calculate each inequality.

1) $x - 3 \leq 7$

2) $8 - x \leq 8$

3) $3x - 9 \leq 9$

4) $4x - 4 \geq 8$

5) $x - 7 \geq 1$

6) $5x - 15 \leq 5$

7) $6x - 8 \leq 4$

8) $-11 + 6x \leq 12$

9) $4(x - 4) \leq 16$

10) $3x - 10 \leq 11$

11) $5x - 25 < 25$

12) $9x - 5 < 22$

13) $20 - 7x \geq -15$

14) $33 + 6x < 45$

15) $8 + 8x \geq 96$

16) $7 + 3x < 13$

17) $4x - 3 < 9$

18) $5(2 - 2x) \geq -30$

19) $-(7 + 6x) < 29$

20) $12 - 8x \geq -20$

21) $-4(x - 6) > 24$

22) $\frac{3x + 9}{6} \leq 10$

23) $\frac{4x - 10}{3} \leq 2$

24) $\frac{2x - 8}{3} > 2$

25) $8 + \frac{x}{6} < 9$

26) $\frac{9x}{7} - 4 < 5$

27) $\frac{15x + 45}{15} > 1$

28) $16 + \frac{x}{4} < 6$

Systems of Equations

Calculate each system of equations.

1) $-x+y=2$ $x=$ ____
$-4x+2y=6$ $y=$ ____

2) $-15x+3y=-9$ $x=$ ____
$9x-16y=48$ $y=$ ____

3) $y=-7$ $x=$ ____
$6x+5y=7$ $y=$ ____

4) $3y=\ -9x+15$ $x=$ ____
$5x-4y=-3$ $y=$ ____

5) $10x-\ 9y=-13$ $x=$ ____
$-5x+3y=11$ $y=$ ____

6) $-12x-16y=20$ $x=$ ____
$6x-12y=30$ $y=$ ____

7) $5x-14y=-23$ $x=$ ____
$-18x+21y=24$ $y=$ ____

8) $15x-21y=-6$ $x=$ ____
$2x-3y=-2$ $y=$ ____

9) $-x+3y=3$ $x=$ ____
$-14x+16y=\ -10$ $y=$ ____

10) $x+5y=50$ $x=$ ____
$3x+10y=80$ $y=$ ____

11) $6x-7y=-8$ $x=$ ____
$-x-\ 4y=-9$ $y=$ ____

12) $2x+4y=-10$ $x=$ ____
$2x-8y=14$ $y=$ ____

13) $4x+3y=12$ $x=$ ____
$5x-3y=15$ $y=$ ____

14) $3x-2y=3$ $x=$ ____
$7x-8y=22$ $y=$ ____

15) $3x+2y=5$ $x=$ ____
$-10x-4y=-14$ $y=$ ____

16) $10x+7y=1$ $x=$ ____
$-5x-7y=24$ $y=$ ____

Systems of Equations Word Problems

Find the answer for each word problem.

1) Tickets to a movie cost $4 for adults and $3 for students. A group of friends purchased 8 tickets for $31.00. How many adults ticket did they buy? ____
2) At a store, Eva bought two shirts and five hats for $77.00. Nicole bought three same shirts and four same hats for $84.00. What is the price of each shirt? _____
3) A farmhouse shelters 18 animals, some are pigs, and some are ducks. Altogether there are 66 legs. How many pigs are there? _____
4) A class of 214 students went on a field trip. They took 36 vehicles, some cars and some buses. If each car holds 5 students and each bus hold 22 students, how many buses did they take? _____
5) A theater is selling tickets for a performance. Mr. Smith purchased 5 senior tickets and 3 child tickets for $105 for his friends and family. Mr. Jackson purchased 3 senior tickets and 5 child tickets for $79. What is the price of a senior ticket? $_____
6) The difference of two numbers is 10. Their sum is 20. What is the bigger number? $_____
7) The sum of the digits of a certain two–digit number is 7. Reversing its digits increase the number by 9. What is the number? _____
8) The difference of two numbers is 11. Their sum is 25. What are the numbers? __________
9) The length of a rectangle is 5 meters greater than 2 times the width. The perimeter of rectangle is 28 meters. What is the length of the rectangle? __________
10) Jim has 25 nickels and dimes totaling $1.80. How many nickels does he have? _____

Answers of Worksheets

One–Step Equations

1) 30	9) 17	17) −14	25) 22
2) 7	10) −4	18) 20	26) 9
3) 4	11) 12	19) 45	27) 60
4) 6	12) 16	20) −13	28) 35
5) 5	13) 34	21) 34	29) 54
6) 11	14) −15	22) −5	30) 24
7) 11	15) −18	23) 47	31) 30
8) 8	16) 14	24) −42	32) −21

Multi–Step Equations

1) 2	9) 2	17) −2	25) 3
2) −7	10) −4	18) −7	26) 14
3) 4	11) −4	19) −5	27) 7
4) 8	12) −5	20) −11	28) 27
5) 6	13) −7	21) −4	29) −5
6) 2	14) 3	22) −2	30) 17
7) 10	15) −2	23) −7	31) −6
8) 2	16) 15	24) 6	32) −4

Graphing Single–Variable Inequalities

1) -1

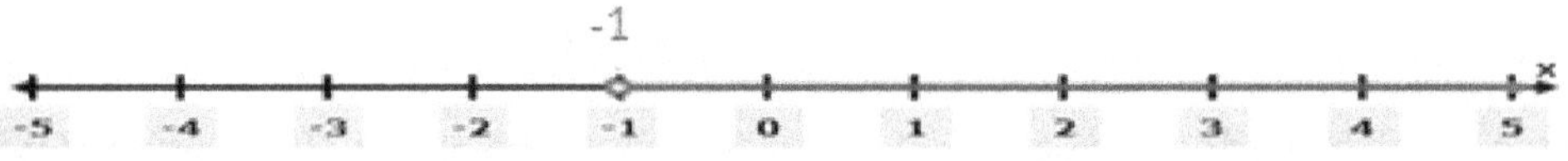

-5 -4 -3 -2 -1 0 1 2 3 4 5 x

2) 2

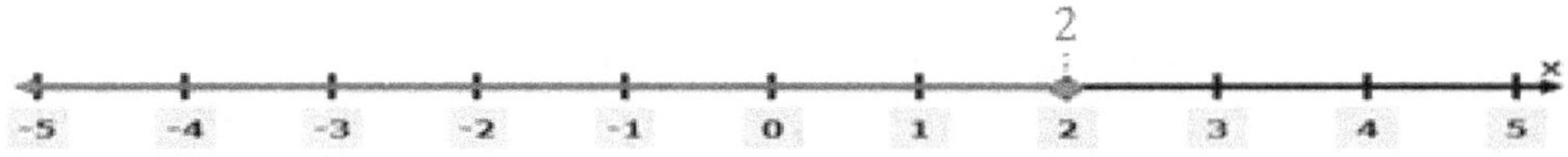

-5 -4 -3 -2 -1 0 1 2 3 4 5 x

3) 0

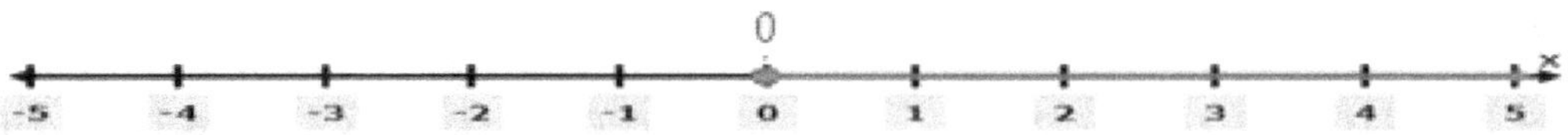

-5 -4 -3 -2 -1 0 1 2 3 4 5 x

4) -3

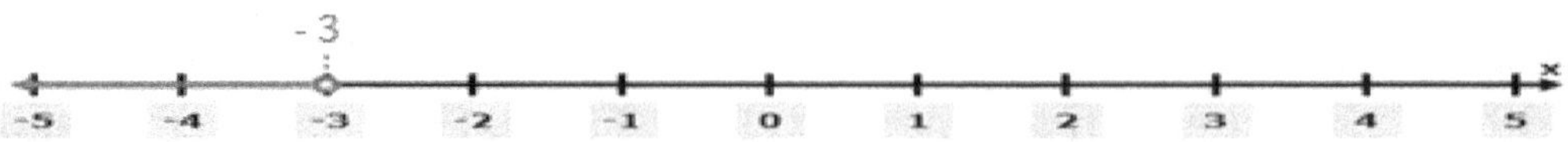

-5 -4 -3 -2 -1 0 1 2 3 4 5 x

5) 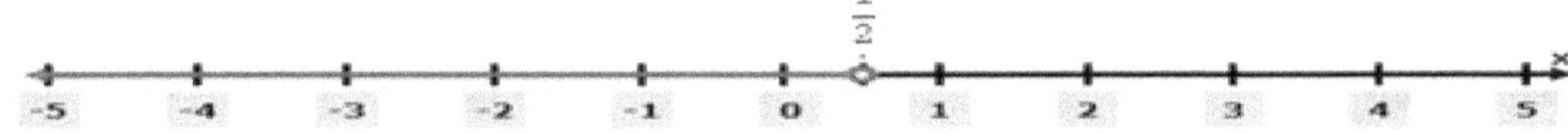

6) 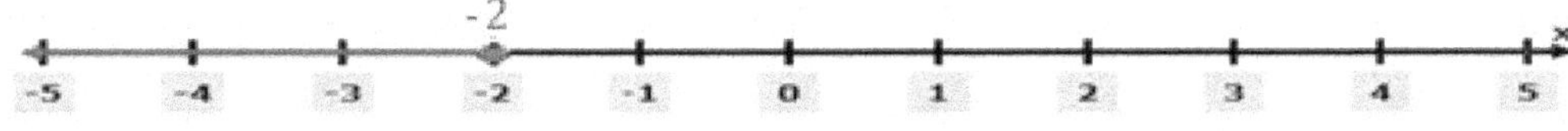

7) 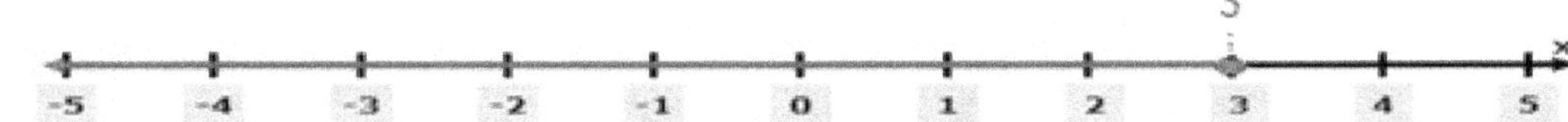

8) 

One–Step Inequalities

1) 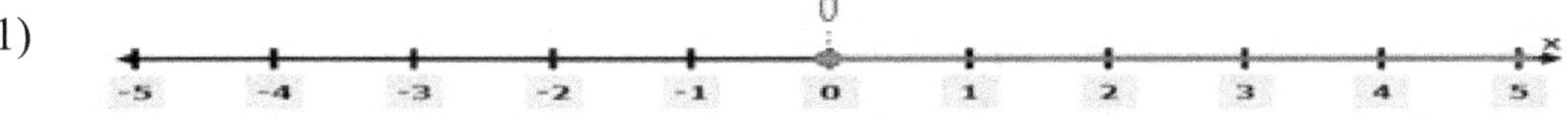

2) 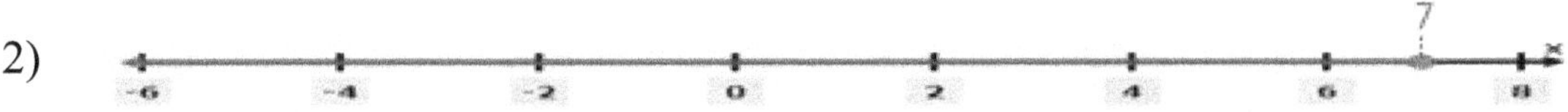

3) 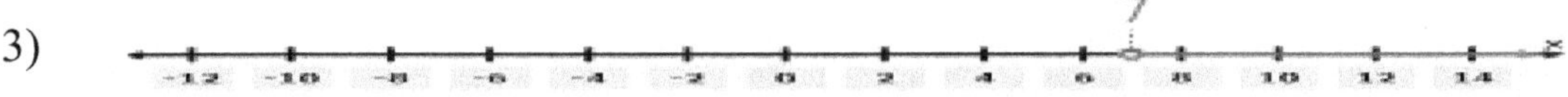

4) 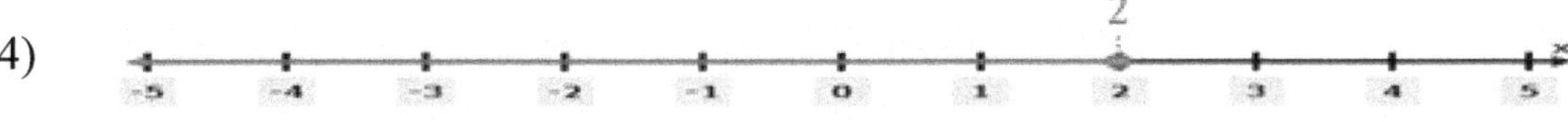

5) 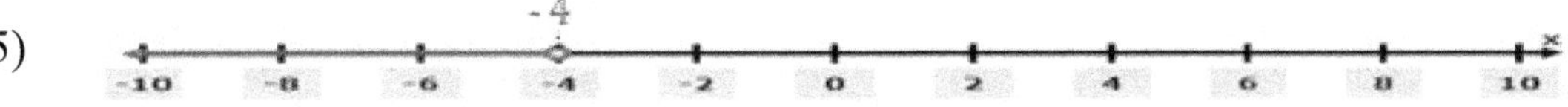

6) 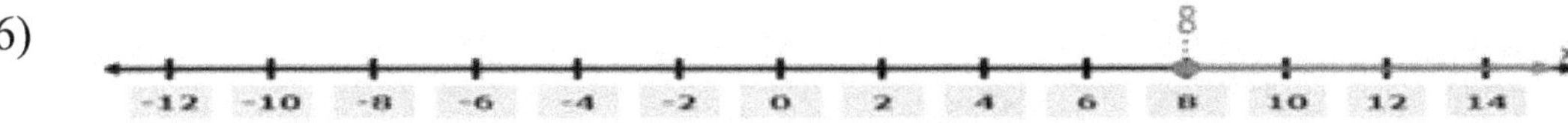

7) 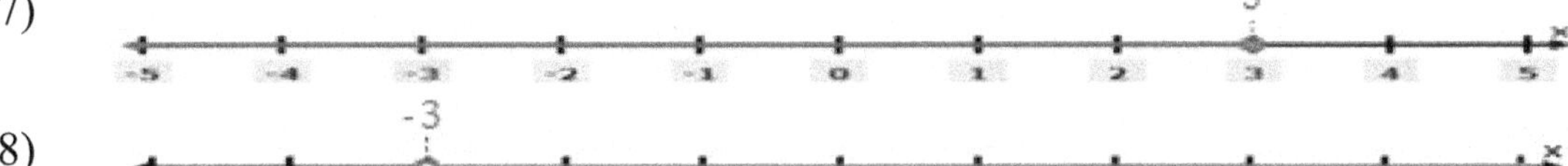

8)
-3
-5 -4 -3 -2 -1 0 1 2 3 4 5

Multi-Step Inequalities

1) $x \leq 10$
2) $x \geq 0$
3) $x \leq 6$
4) $x \geq 3$
5) $x \geq 8$
6) $x \leq 4$
7) $x \leq 2$
8) $x \leq \frac{23}{6}$
9) $x \leq 8$
10) $x \leq 7$
11) $x < 10$
12) $x < 3$
13) $x \leq 5$
14) $x < 2$
15) $x \geq 11$
16) $x < 2$
17) $x < 3$
18) $x \leq 4$
19) $x > -6$
20) $x \leq 4$
21) $x < 0$
22) $x \leq 17$
23) $x \leq 4$
24) $x > 7$

25) $x < 6$

26) $x < 7$

27) $x > -2$

28) $x < -40$

Systems of Equations

1) $x = -1, y = 1$
2) $x = 0, y = -3$
3) $x = 7$
4) $x = 1, y = 2$
5) $x = -4, y = -3$
6) $x = 1, y = -2$
7) $x = 1, y = 2$
8) $x = 8, y = 6$
9) $x = 3, y = 2$
10) $x = -20, y = 14$
11) $x = 1, y = 2$
12) $x = -1, y = -2$
13) $x = 3, y = 0$
14) $x = -2, y = -\frac{9}{2}$
15) $x = 1, y = 1$
16) $x = 5, y = -7$

Systems of Equations Word Problems

1) 7
2) \$16
3) 15
4) 2
5) \$18
6) 15
7) 34
8) 18, 7
9) 11 meters
10) 14

Chapter 7 :

Linear Functions

Topics that you'll practice in this chapter:

- ✓ Finding Slope
- ✓ Graphing Lines Using Line Equation
- ✓ Writing Linear Equations
- ✓ Graphing Linear Inequalities
- ✓ Finding Midpoint
- ✓ Finding Distance of Two Points

Life is not linear; you have ups and downs. It's how you deal with the troughs that defines you.

Michael Lee-Chin

Finding Slope

Find the slope of each line.

1) $y = x + 8$

2) $y = -3x + 5$

3) $y = 2x + 12$

4) $y = -4x + 19$

5) $y = 11 + 6x$

6) $y = 7 - 5x$

7) $y = 8x + 19$

8) $y = -9x + 20$

9) $y = -7x + 4$

10) $y = 3x - 8$

11) $y = \frac{1}{3}x + 8$

12) $y = -\frac{4}{5}x + 9$

13) $-3x + 6y = 30$

14) $4x + 4y = 16$

15) $3y - x = 10$

16) $8y - x = 5$

Find the slope of the line through each pair of points.

17) $(2, 3), (7, 10)$

18) $(-3,5), (2, 15)$

19) $(5, -3), (1,9)$

20) $(-5, -5), (10, 25)$

21) $(22, 3), (7, 18)$

22) $(-16, 8), (-7, 26)$

23) $(25, 11), (29, 19)$

24) $(26, -19), (14, 17)$

25) $(22, -13), (20, -11)$

26) $(19, 7), (15, -3)$

27) $(5, 7), (11, 19)$

28) $(52, -62), (40, 70)$

Graphing Lines Using Line Equation

Sketch the graph of each line.

1) $y = x - 2$

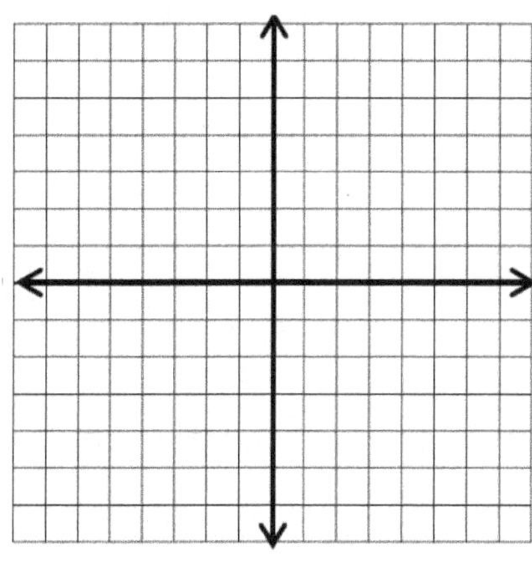

2) $y = -3x + 2$

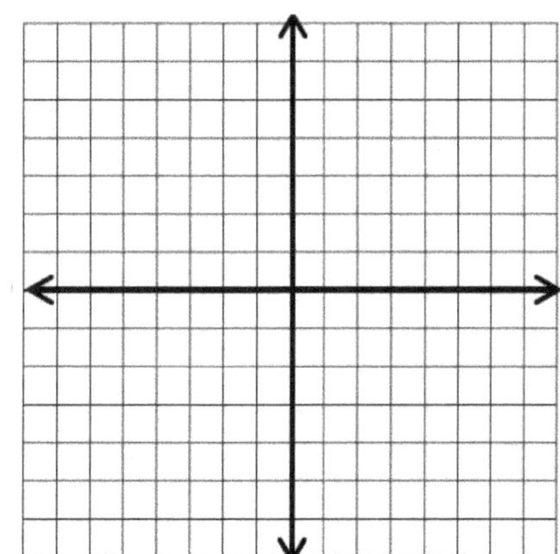

3) $x + y = 0$

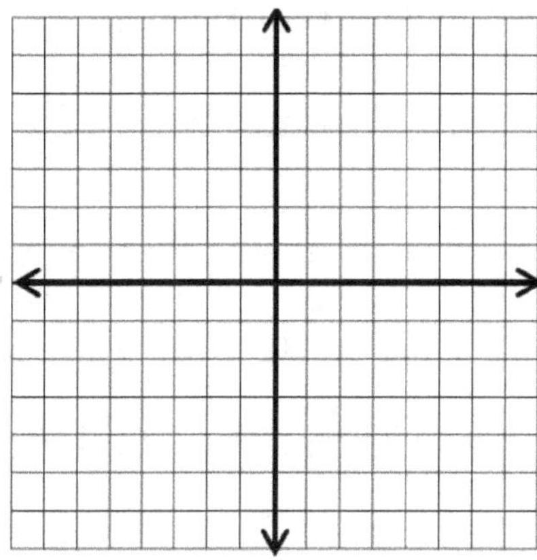

4) $x + y = -3$

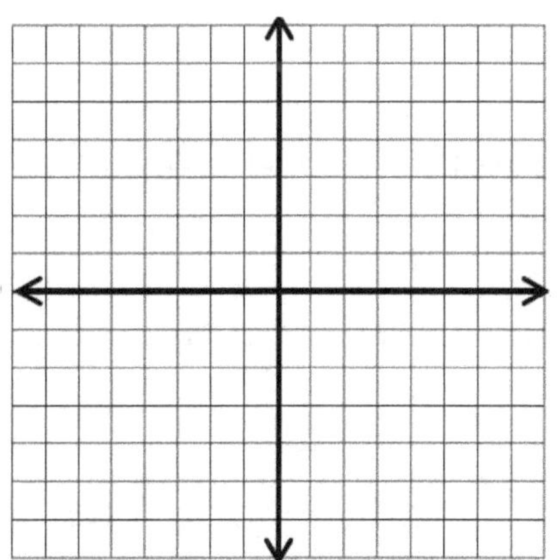

5) $2x + 3y = -4$

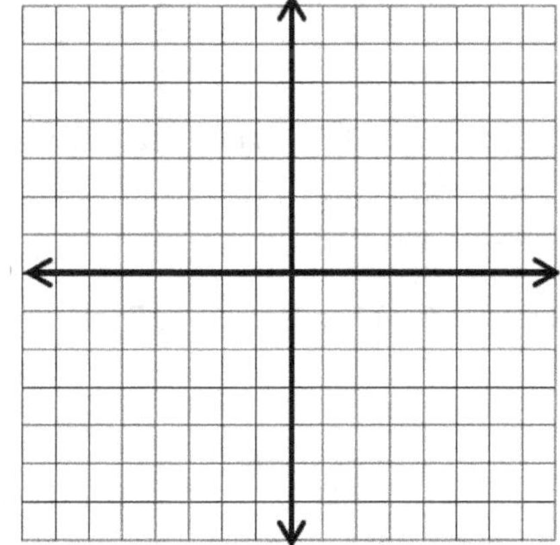

6) $y - 3x + 6 = 0$

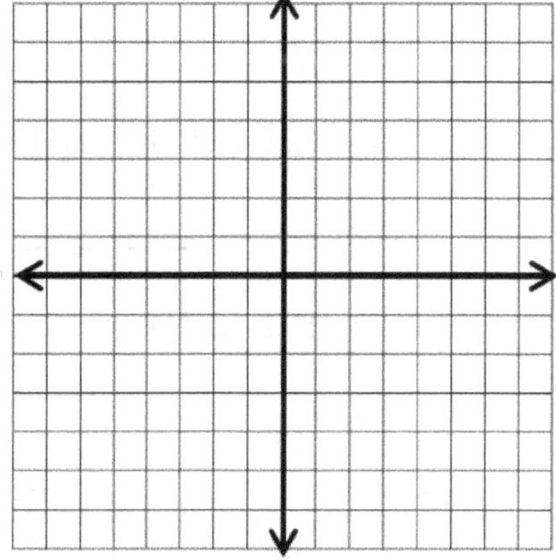

Writing Linear Equations

Write the equation of the line through the given points.

1) Through: $(2,-5),(3,9)$

2) Through: $(-6,3),(3,12)$

3) Through: $(10,7),(5,27)$

4) Through: $(15,11),(3,-1)$

5) Through: $(24,17),(12,-7)$

6) Through: $(8,29),(4,-7)$

7) Through: $(20,-16),(12,0)$

8) Through: $(-3,10),(2,-5)$

9) Through: $(-6,17),(4,-3)$

10) Through: $(-8,22),(5,-4)$

11) Through: $(9,27),(3,-3)$

12) Through: $(11,32),(9,4)$

13) Through: $(-3,13),(-4,0)$

14) Through: $(-5,5),(5,15)$

15) Through: $(18,-32),(11,3)$

16) Through: $(-4,25),(4,-15)$

Find the answer for each problem.

17) What is the equation of a line with slope 6 and intercept 12? _______________

18) What is the equation of a line with slope -11 and intercept -4? _______________

19) What is the equation of a line with slope -3 and passes through point $(5,2)$? _______________

20) What is the equation of a line with slope -5 and passes through point $(-2,-1)$? _______________

21) The slope of a line is -10 and it passes through point $(-3,0)$. What is the equation of the line? _______________

22) The slope of a line is 8 and it passes through point $(0,7)$. What is the equation of the line? _______________

Graphing Linear Inequalities

Sketch the graph of each linear inequality.

1) $y > 4x - 5$ 2) $y < 2x + 4$ 3) $y \leq -5x - 2$

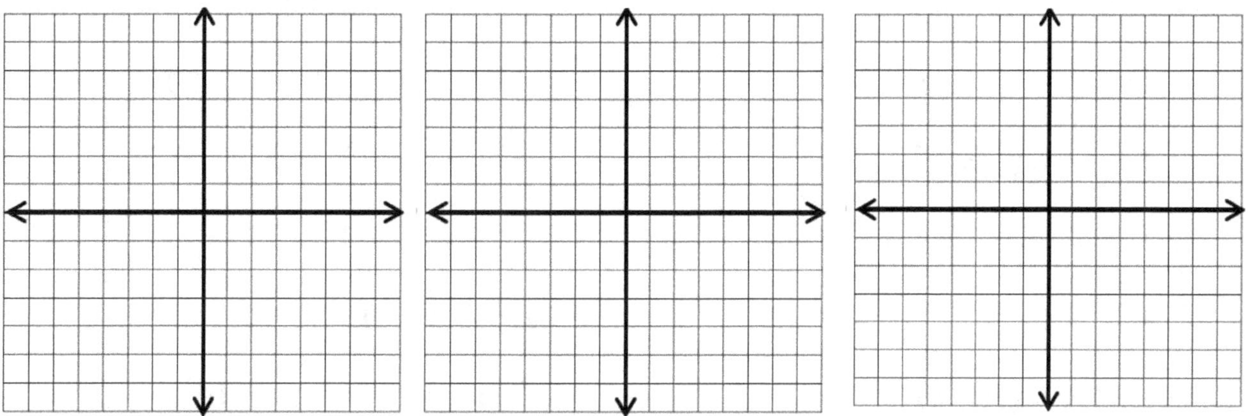

4) $4y \geq 12 + 4x$ 5) $-12y < 3x - 24$ 6) $5y \geq -15x + 10$

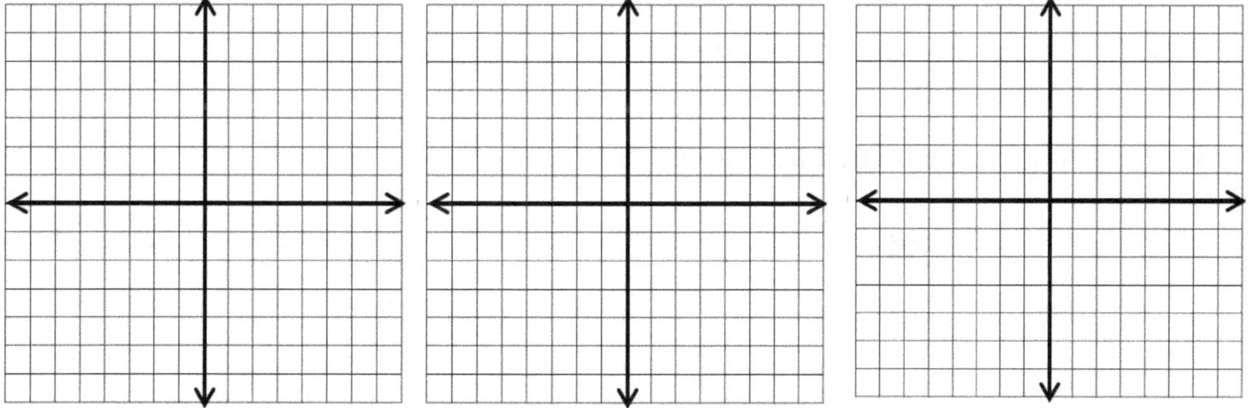

Finding Midpoint

Find the midpoint of the line segment with the given endpoints.

1) $(-4,-3),(2,3)$
2) $(9,0),(-1,8)$
3) $(9,-6),(3,14)$
4) $(-10,-6),(0,8)$
5) $(2,-5),(14,-15)$
6) $(-10,-3),(4,-13)$
7) $(8,7),(-8,13)$
8) $(-3,6),(-9,2)$
9) $(-4,5),(16,-9)$
10) $(7,14),(9,-2)$
11) $(-8,6),(6,6)$
12) $(10,5),(-2,-3)$
13) $(-5,12),(-3,3)$
14) $(12,7),(8,-2)$
15) $(10,2),(-6,14)$
16) $(-1,-2),(-7,10)$
17) $(7,-7),(13,-13)$
18) $(-3,-8),(11,-4)$
19) $(5,-11),(-8,9)$
20) $(14,-4),(16,14)$
21) $(0,-5),(8,-1)$
22) $(3,0),(-21,18)$
23) $(17,-3),(-7,-5)$
24) $(26,-12),(6,24)$

Find the answer for each problem.

25) One endpoint of a line segment is $(-3,7)$ and the midpoint of the line segment is $(-6,9)$. What is the other endpoint? ______________

26) One endpoint of a line segment is $(-3,7)$ and the midpoint of the line segment is $(1,5)$. What is the other endpoint? ______________

27) One endpoint of a line segment is $(-10,-16)$ and the midpoint of the line segment is $(2,9)$. What is the other endpoint? ______________

Finding Distance of Two Points

Find the distance between each pair of points.

1) $(6,3),(-3,-9)$

2) $(5,2),(-10,-6)$

3) $(8,5),(8,3)$

4) $(-8,-2),(2,22)$

5) $(6,-7),(-3,-7)$

6) $(12,0),(-9,-20)$

7) $(3,20),(3,-5)$

8) $(10,17),(5,5)$

9) $(7,-2),(-4,-2)$

10) $(13,4),(5,-2)$

11) $(11,13),(5,5)$

12) $(1,4),(-23,-3)$

13) $(9,8),(5,-4)$

14) $(-11,-4),(5,8)$

15) $(-2,-6),(-2,-12)$

16) $(-1,-4),(23,3)$

17) $(19,3),(7,-6)$

18) $(-5,-2),(3,4)$

19) $(2,6),(2,-12)$

20) $(-4,-2),(8,-2)$

Find the answer for each problem.

21) Triangle ABC is a right triangle on the coordinate system and its vertices are $(-2,5)$, $(-2,1)$, and $(1,1)$. What is the area of triangle ABC? ______________

22) Three vertices of a triangle on a coordinate system are $(3,-6)$, $(-5,-12)$, and $(3,-18)$. What is the perimeter of the triangle? ________

23) Four vertices of a rectangle on a coordinate system are $(-2,2)$, $(-2,6)$, $(4,2)$, and $(4,6)$. What is its perimeter? ______________

Answers of Worksheets

Finding Slope

1) 1
2) -3
3) 2
4) -4
5) 6
6) -5
7) 8
8) -9
9) -7
10) 3
11) $\frac{1}{3}$
12) $-\frac{4}{5}$
13) $\frac{1}{2}$
14) -1
15) $\frac{1}{3}$
16) $\frac{1}{8}$
17) $\frac{7}{5}$
18) 2
19) -3
20) 2
21) -1
22) 2
23) 2
24) -3
25) -1
26) $\frac{5}{2}$
27) 2
28) -11

Graphing Lines Using Line Equation

1) $y = x - 2$

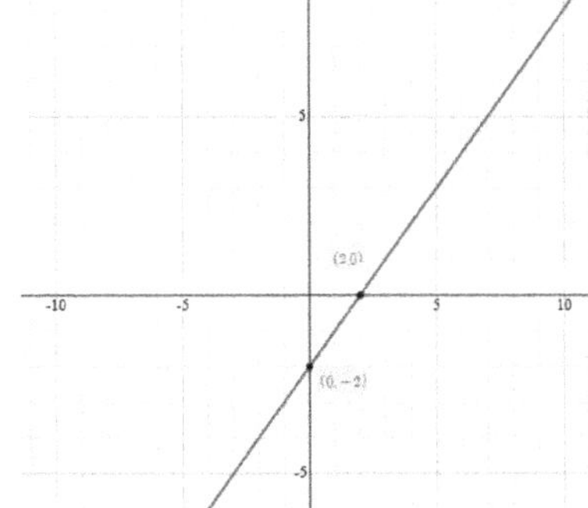

2) $y = -3x + 2$

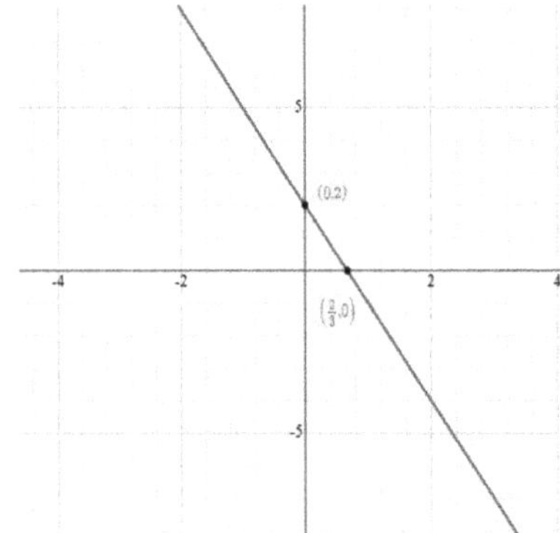

3) $x + y = 0$

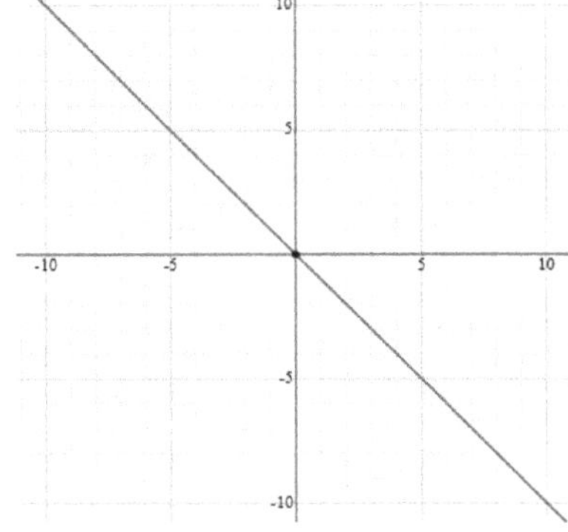

4) $x + y = -3$

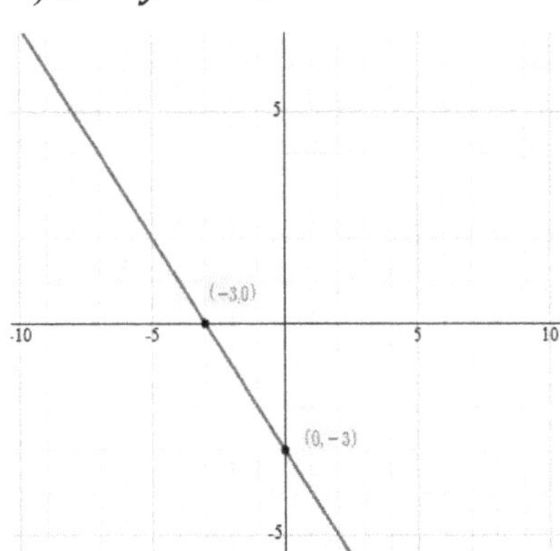

5) $2x + 3y = -4$

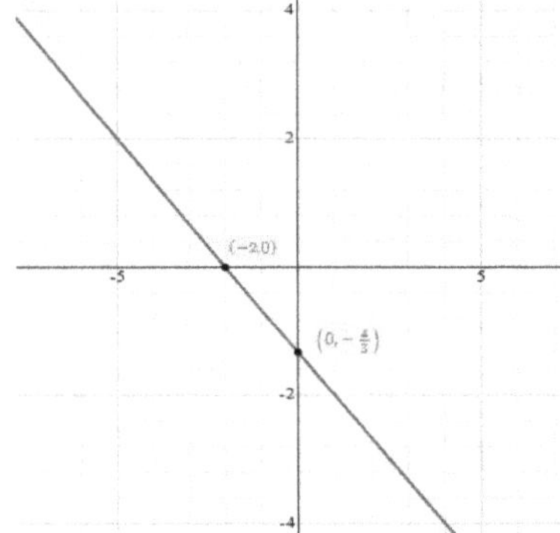

6) $y - 3x + 6 = 0$

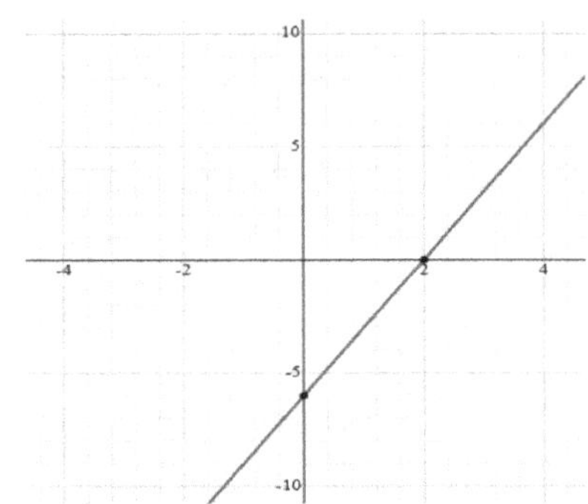

Writing Linear Equations

1) $y = 14x - 33$
2) $y = x + 9$
3) $y = -4x + 47$
4) $y = x - 4$
5) $y = 2x - 31$
6) $y = 9x - 43$
7) $y = -2x + 24$
8) $y = -3x + 1$
9) $y = -2x + 5$
10) $y = -2x + 6$
11) $y = 5x - 18$
12) $y = 14x - 122$
13) $y = 13x + 52$
14) $y = x + 10$
15) $y = -5x + 58$
16) $y = -5x + 5$
17) $y = 6x + 12$
18) $y = -11x - 4$
19) $y = -3x + 17$
20) $y = -5x - 11$
21) $y = -10x - 30$
22) $y = 8x + 7$

Graphing Linear Inequalities

1) $y > 4x - 5$

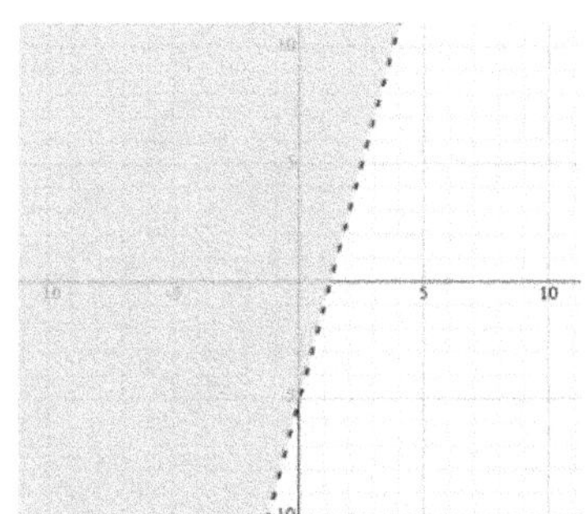

2) $y < 2x + 4$

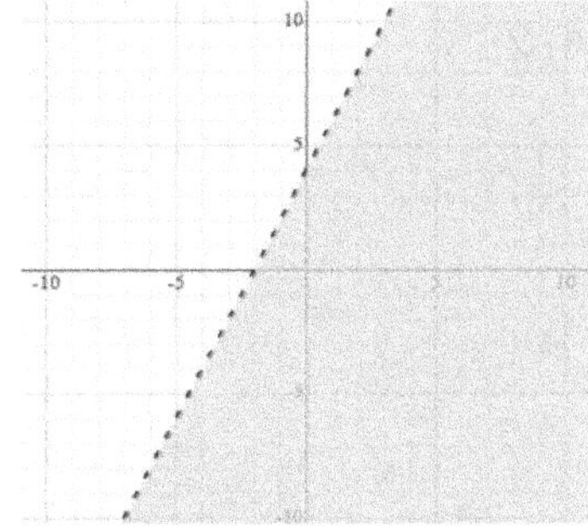

3) $y \leq -5x - 2$

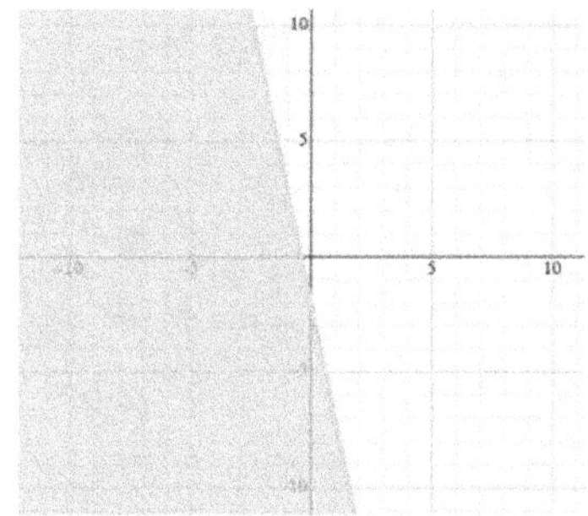

4) $4y \geq 12 + 4x$

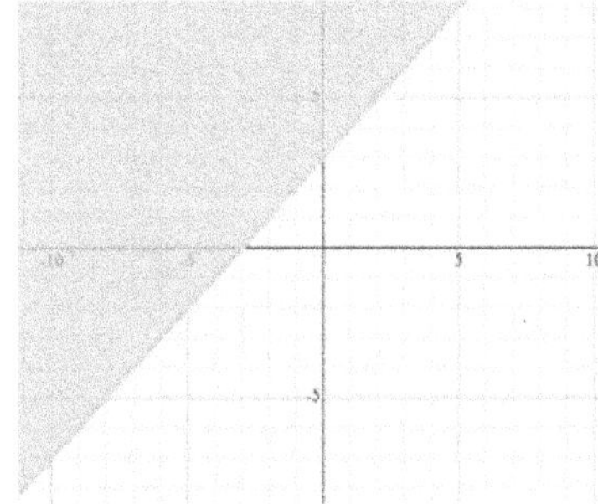

5) $-12y < 3x - 24$

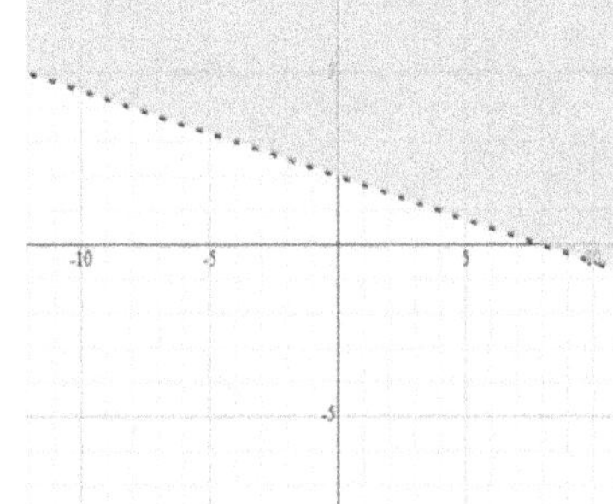

6) $5y \geq -15x + 10$

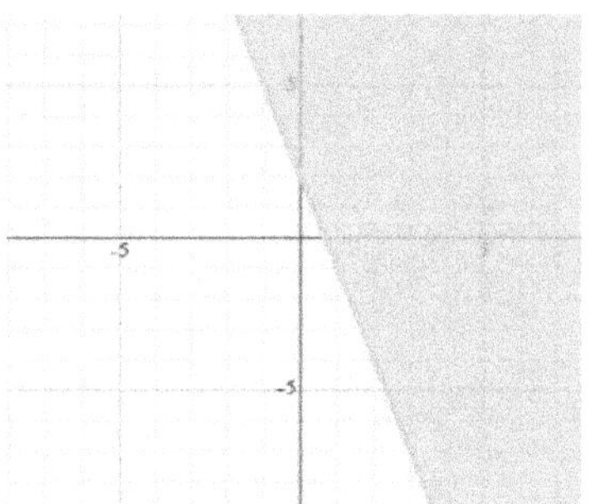

Finding Midpoint

1) $(-1, 0)$
2) $(4, 4)$
3) $(6, 4)$
4) $(-5, 1)$
5) $(8, -10)$
6) $(-3, -8)$
7) $(0, 10)$
8) $(-6, 4)$
9) $(6, -2)$
10) $(8, 6)$
11) $(-1, 6)$
12) $(4, 1)$
13) $(-4, 7.5)$
14) $(10, 2.5)$
15) $(2, 8)$
16) $(-4, 4)$
17) $(10, -10)$
18) $(4, -6)$

19) $(-1.5, -1)$
20) $(15, 5)$
21) $(4, -3)$
22) $(-9, 9)$
23) $(5, -4)$
24) $(16, 6)$
25) $(-9, 11)$
26) $(5, 3)$
27) $(14, 34)$

Finding Distance of Two Points

1) 15
2) 17
3) 2
4) 26
5) 9
6) 29
7) 25
8) 13
9) 11
10) 10
11) 10
12) 25
13) $4\sqrt{10}$
14) 20
15) 6
16) 25
17) 15
18) 10
19) 18
20) 12
21) 6 *square units*
22) 32 *units*
23) 20 *units*

Chapter 8 :
Polynomials

Topics that you'll practice in this chapter:

- ✓ Writing Polynomials in Standard Form
- ✓ Simplifying Polynomials
- ✓ Adding and Subtracting Polynomials
- ✓ Multiplying Monomials
- ✓ Multiplying and Dividing Monomials
- ✓ Multiplying a Polynomial and a Monomial
- ✓ Multiplying Binomials
- ✓ Factoring Trinomials
- ✓ Operations with Polynomials

Mathematics is the supreme judge; from its decisions there is no appeal.

– Tobias Dantzig

Writing Polynomials in Standard Form

Write each polynomial in standard form.

1) $11x - 7x =$

2) $-5 + 19x - 19x =$

3) $6x^5 - 12x^3 =$

4) $12 + 17x^4 - 12 =$

5) $5x^2 + 4x - 9x^3 =$

6) $-3x^2 + 12x^5 =$

7) $5x + 8x^3 - 2x^8 =$

8) $-7x^3 + 4x - 9x^6 =$

9) $3x^2 + 22 - 6x =$

10) $3 - 4x + 9x^4 =$

11) $13x^2 + 28x - 8x^3 =$

12) $16 + 4x^2 - 2x^3 =$

13) $19x^2 - 9x + 9x^4 =$

14) $3x^4 - 7x^2 - 2x^3 =$

15) $-51 + 3x^2 - 8x^4 =$

16) $7x^2 - 8x^6 + 4x^4 - 15 =$

17) $6x^4 - 4x^5 + 16 - 3x^3 =$

18) $-2x^6 + 4x - 7x^2 - 5x =$

19) $11x^7 + 8x^5 - 5x^7 - 3x^2 =$

20) $2x^2 - 12x^5 + 8x^2 + 3x^6 =$

21) $4x^5 - 11x^7 - 6x^3 + 16x^5 =$

22) $6x^3 + 3x^5 + 34x^4 - 8x^5 =$

23) $3x(4x + 5 - 2x^2) =$

24) $12x(x^6 + 4x^3) =$

25) $5x(3x^2 + 6x + 4) =$

26) $7x(4 - 2x + 6x^5) =$

27) $3x(4x^4 - 4x^3 + 2) =$

28) $4x(2x^5 + 6x^2 - 3) =$

29) $5x(3x^4 + 4x^3 + 2x) =$

30) $2x(3x - 2x^3 + 4x^6) =$

Simplifying Polynomials

Simplify each expression.

1) $3(4x - 20) =$

2) $5x(3x - 4) =$

3) $6x(5x - 7) =$

4) $3x(7x + 5) =$

5) $5x(4x - 3) =$

6) $6x(8x + 2) =$

7) $(3x - 2)(x - 4) =$

8) $(x - 5)(2x + 6) =$

9) $(x - 3)(x - 7) =$

10) $(3x + 4)(3x - 4) =$

11) $(5x - 4)(5x - 2) =$

12) $6x^2 + 6x^2 - 8x^4 =$

13) $3x - 2x^2 + 5x^3 + 7 =$

14) $7x + 4x^2 - 10x^3 =$

15) $12x^2 + 5x^5 - 6x^3 =$

16) $-5x^2 + 4x^6 + 6x^8 =$

17) $-12x^3 + 10x^5 - 4x^6 + 4x =$

18) $11 - 7x^2 + 4x^2 - 16x^3 + 11 =$

19) $2x^2 - 9x + 4x^3 + 15x - 10x =$

20) $13 - 7x^5 + 6x^5 - 4x^2 + 5 =$

21) $-5x^8 + x^6 - 14x^3 + 5x^8 =$

22) $(7x^4 - 4) + (7x^4 - 2x^4) =$

23) $3(3x^4 - 4x^3 - 6x^4) =$

24) $-5(x^9 + 8) - 5(10 - x^9) =$

25) $8x^3 - 9x^4 - 2x + 19 - 8x^3 =$

26) $11 - 8x^3 + 6x^3 - 7x^5 + 6 =$

27) $(5x^3 - 4x) - (6x - 2 - 6x^3) =$

28) $4x^2 - 5x^4 - x(3x^3 + 2x) =$

29) $6x + 6x^5 - 10 - 4(x^5 - 3) =$

30) $4 - 3x^4 + (6x^5 - 2x^4 + 5x^5) =$

31) $-(x^5 + 4) - 8(3 + x^5) =$

32) $(4x^3 - 3x) - (3x - 5x^3) =$

Adding and Subtracting Polynomials

Add or subtract expressions.

1) $(-2x^2 - 3) + (3x^2 + 4) =$

2) $(4x^3 + 6) - (7 - 2x^3) =$

3) $(4x^5 + 5x^2) - (2x^5 + 15) =$

4) $(6x^3 - 2x^2) + (5x^2 - 4x) =$

5) $(10x^4 + 28x) - (34x^4 + 6) =$

6) $(7x^2 - 3) + (7x^2 + 3) =$

7) $(9x^2 + 4) - (10 - 5x^2) =$

8) $(6x^2 + x^5) - (x^5 + 4) =$

9) $(4x^3 - x) + (3x - 7x^3) =$

10) $(11x + 10) - (8x + 10) =$

11) $(15x^3 - 3x) - (3x - 4x^3) =$

12) $(4x - x^5) - (6x^5 + 8x) =$

13) $(2x^2 - 7x^7) - (4x^7 - 6x) =$

14) $(3x^2 - 5) + (8x^2 + 4x^5) =$

15) $(9x^4 + 5x^5) - (x^5 - 9x^4) =$

16) $(-4x^3 - 2x) + (9x - 5x^3) =$

17) $(4x - 3x^2) - (148x^2 + x) =$

18) $(5x - 8x^4) - (3x^4 - 4x^2) =$

19) $(8x^4 - 4) + (2x^4 - 3x^2) =$

20) $(5x^6 + 7x^3) - (x^3 - 5x^6) =$

21) $(-2x^2 + 20x^5 + 5x^4) + (12x^4 + 8x^5 + 24x^2) =$

22) $(7x^4 - 9x^7 - 6x) - (-3x^4 - 9x^7 + 6x) =$

23) $(14x + 12x^4 - 18x^6) + (20x^4 + 18x^6 - 10x) =$

24) $(5x^8 - 6x^6 - 4x) - (5x^3 + 9x^6 - 7x) =$

25) $(11x^2 - 6x^4 - 3x) - (-4x^2 - 12x^4 + 9x) =$

26) $(-5x^9 + 14x^3 + 3x^7) + (10x^7 + 26x^3 + 3x^9) =$

Multiplying Monomials

Simplify each expression.

1) $6u^8 \times (-u^2) =$

2) $(-5p^8) \times (-2p^3) =$

3) $4xy^3z^5 \times 3z^4 =$

4) $3u^5t \times 8ut^4 =$

5) $(-5a^2) \times (-7a^3b^6) =$

6) $-3a^4b^3 \times 6a^2b =$

7) $13xy^5 \times x^4y^4 =$

8) $6p^4q^3 \times (-8pq^6) =$

9) $8s^4t^3 \times 4st^3 =$

10) $(-6x^4y^3) \times 6x^2y =$

11) $3xy^7z \times 12z^3 =$

12) $24xy \times x^2y =$

13) $13pq^4 \times (-3p^2q) =$

14) $13s^3t^4 \times st^4 =$

15) $11p^5 \times (-6p^3) =$

16) $(-8p^3q^5r) \times 3pq^4r^6 =$

17) $(-4a^4) \times (-7a^3b) =$

18) $6u^6v^2 \times (-5u^3v^4) =$

19) $9u^5 \times (-3u) =$

20) $-6xy^5 \times 4x^2y =$

21) $13y^5z^3 \times (-y^3z) =$

22) $8a^4bc^3 \times 2abc^3 =$

23) $(-7p^5q^6) \times (-5p^4q^2) =$

24) $4u^5v^3 \times (-4u^7v^3) =$

25) $17y^4z^5 \times (-y^6z) =$

26) $(-5pq^3r^2) \times 8p^2q^4r =$

27) $3ab^5c^6 \times 5a^4bc^2 =$

28) $6x^3yz^2 \times 3x^2y^7z^3 =$

Multiplying and Dividing Monomials

Simplify each expression.

1) $(5x^5)(2x^2) =$

2) $(4x^4)(6x^2) =$

3) $(3x^4)(7x^4) =$

4) $(5x^6)(4x^2) =$

5) $(12x^4)(3x^6) =$

6) $(4yx^8)(8y^4x^3) =$

7) $(14\text{x}^4y)(x^3y^5) =$

8) $(-5x^3y^4)(2x^3y^5) =$

9) $(-6x^4y^2)(-3x^3y^5) =$

10) $(5x^3y)(-5x^2y^3) =$

11) $(6x^4y^3)(4x^3y^4) =$

12) $(4x^3y^2)(5x^2y^4) =$

13) $(12x^3y^6)(4x^4y^{10}) =$

14) $(15x^3y^5)(3x^4y^6) =$

15) $(7x^2y^7)(8x^6y^7) =$

16) $(-3x^3y^8)(7x^9y^4) =$

17) $\frac{5x^6y^6}{xy^4} =$

18) $\frac{19x^7y^5}{19x^6y} =$

19) $\frac{56x^4y^4}{8xy} =$

20) $\frac{81x^5y^6}{9x^4y^5} =$

21) $\frac{36x^7y^6}{9x^2y^3} =$

22) $\frac{48x^9y^7}{4x^4y^6} =$

23) $\frac{88x^{18}y^{12}}{11x^8y^9} =$

24) $\frac{30x^7y^6}{6x^8y^3} =$

25) $\frac{150x^7y^6}{30x^4y^6} =$

26) $\frac{-42x^{18}y^{14}}{6x^4y^9} =$

27) $\frac{-36x^7y^8}{9x^5y^8} =$

Multiplying a Polynomial and a Monomial

Find each product.

1) $x(2x+4)=$

2) $6(4-2x)=$

3) $5x(4x+2)=$

4) $x(-4x+5)=$

5) $8x(2x-2)=$

6) $6(2x-4y)=$

7) $7x(5x-5)=$

8) $3x(12x+2y)=$

9) $4x(x+6y)=$

10) $11x(3x+4y)=$

11) $7x(3x+2)=$

12) $10x(4x-10y)=$

13) $9x(3x-2y)=$

14) $7x(x-4y+6)=$

15) $8x(2x^2+5y^2)=$

16) $12x(2x+3y)=$

17) $4(2x^4-4y^4)=$

18) $4x(-3x^2y+4y)=$

19) $-4(5x^3-2xy+4)=$

20) $4(x^2-5xy-6)=$

21) $8x(2x^3-5xy+2x)=$

22) $-6x(-2x^3-6x+2xy)=$

23) $3(2x^2+xy-9y^2)=$

24) $4x(5x^3-3x+7)=$

25) $6(3x^{22}-2x-5)=$

26) $x^2(-2x^3+4x+3)=$

27) $x^2(4x^3+10-2x)=$

28) $4x^4(3x^3-2x+5)=$

29) $2x^2(4x^4-5xy+7y^3)=$

30) $5x^2(5x^4-3x+9)=$

31) $7x^2(6x^2+3x-6)=$

32) $4x(x^3-4xy+2y^2)=$

Multiplying Binomials

Find each product.

1) $(x+3)(x+6)=$

2) $(x-4)(x+3)=$

3) $(x-3)(x-8)=$

4) $(x+8)(x+9)=$

5) $(x-2)(x-12)=$

6) $(x+5)(x+5)=$

7) $(x-6)(x+7)=$

8) $(x-8)(x-3)=$

9) $(x+7)(x+12)=$

10) $(x-4)(x+8)=$

11) $(x+8)(x+8)=$

12) $(x+2)(x+7)=$

13) $(x-6)(x+6)=$

14) $(x-5)(x+5)=$

15) $(x+11)(x+11)=$

16) $(x+6)(x+9)=$

17) $(x-2)(x+2)=$

18) $(x-4)(x+7)=$

19) $(3x+5)(x+6)=$

20) $(5x-6)(4x+8)=$

21) $(x-7)(3x+7)=$

22) $(x-9)(x-4)=$

23) $(x-12)(x+2)=$

24) $(2x-4)(5x+4)=$

25) $(3x-8)(x+8)=$

26) $(7x-2)(6x+3)=$

27) $(4x+5)(3x+5)=$

28) $(7x-4)(9x+4)=$

29) $(x+2)(2x-8)=$

30) $(5x-4)(5x+4)=$

31) $(3x+2)(3x-7)=$

32) $(x^2+8)(x^2-8)=$

Factoring Trinomials

Factor each trinomial.

1) $x^2 + 8x + 12 =$
2) $x^2 - 6x + 5 =$
3) $x^2 + 15x + 36 =$
4) $x^2 - 12x + 35 =$
5) $x^2 - 11x + 18 =$
6) $x^2 - 9x + 18 =$
7) $x^2 + 18x + 72 =$
8) $x^2 - x - 72 =$
9) $x^2 + 4x - 21 =$
10) $x^2 - 13x + 22 =$
11) $x^2 + 2x - 24 =$
12) $x^2 - 3x - 40 =$
13) $x^2 - 3x - 70 =$
14) $x^2 + 26x + 169 =$
15) $4x^2 - 7x - 15 =$
16) $x^2 - 14x + 33 =$
17) $10x^2 + 5x - 15 =$
18) $6x^2 - 4x - 42 =$
19) $x^2 + 12x + 36 =$
20) $5x^2 + 17x - 12 =$

Calculate each problem.

21) The area of a rectangle is $x^2 - x - 56$. If the width of rectangle is $x + 7$, what is its length? ______________

22) The area of a parallelogram is $4x^2 + 17x - 15$ and its height is $x + 5$. What is the base of the parallelogram? ______________

23) The area of a rectangle is $6x^2 - 22x + 12$. If the width of the rectangle is $3x - 2$, what is its length? ______________

Operations with Polynomials

Find each product.

1) $4(5x + 3) =$ ____________

2) $8(2x + 6) =$ ____________

3) $2(5x - 2) =$ ____________

4) $-4(7x - 3) =$ ____________

5) $3x^2(9x + 1) =$ ____________

6) $4x^6(7x - 9) =$ ____________

7) $3x^4(-7x + 3) =$ ____________

8) $-8x^4\ (5x - 8) =$ ____________

9) $7\ (x^2 + 5x - 3) =$ ____________

10) $9(5x^2 - 7x + 5) =$ ____________

11) $3(3x^2 + 3x + 2) =$ ____________

12) $5x(3x^2 + 5x + 8) =$ ____________

13) $(5x + 7)(3x - 3) =$ ____________

14) $(9x + 3)(3x - 5) =$ ____________

15) $(6x + 3)(4x - 2) =$ ____________

16) $(7x - 2)(3x + 5) =$ ____________

Calculate each problem.

17) The measures of two sides of a triangle are $(2x + 5y)$ and $(6x - 3y)$. If the perimeter of the triangle is $(13x + 4y)$, what is the measure of the third side? ____________

18) The height of a triangle is $(8x + 5)$ and its base is $(4x - 3)$. What is the area of the triangle? ____________

19) One side of a square is $(6x + 2)$. What is the area of the square? ____________

20) The length of a rectangle is $(5x - 8y)$ and its width is $(15x + 8y)$. What is the perimeter of the rectangle? ____________

21) The side of a cube measures $(x + 2)$. What is the volume of the cube? ____________

22) If the perimeter of a rectangle is $(28x + 6y)$ and its width is $(5x + 2y)$, what is the length of the rectangle? ____________

Answers of Worksheets

Writing Polynomials in Standard Form

1) $4x$
2) -5
3) $6x^5 - 12x^3$
4) $14x^4$
5) $-9x^3 + 5x^2 + 4x$
6) $12x^5 - 3x^2$
7) $-2x^8 + 8x^3 + 5x$
8) $-9x^6 - 7x^3 + 4x$
9) $3x^2 - 6x + 22$
10) $9x^4 - 4x + 3$
11) $-8x^3 + 13x^2 + 28x$
12) $-2x^3 + 4x^2 + 16$
13) $9x^4 + 19x^2 - 9x$
14) $3x^4 - 2x^3 - 7x^2$
15) $-8x^4 + 3x^2 - 51$
16) $-8x^6 + 4x^4 + 7x^2 - 15$
17) $-4x^5 + 6x^4 - 3x^3 + 16$
18) $-2x^6 - 7x^2 - x$
19) $6x^7 + 8x^5 - 3x^2$
20) $3x^6 - 12x^5 + 10x^2$
21) $-11x^7 + 20x^5 - 6x^3$
22) $-5x^5 + 34x^4 + 6x^3$
23) $-6x^3 + 12x^2 + 15x$
24) $12x^7 + 48x^4$
25) $15x^3 + 30x^2 + 20x$
26) $42x^6 - 14x^2 + 28x$
27) $12x^5 - 12x^4 + 6x$
28) $8x^6 + 24x^3 - 12x$
29) $15x^5 + 20x^4 + 10x^2$
30) $8x^7 - 4x^4 + 6x^2$

Simplifying Polynomials

1) $12x - 60$
2) $15x^2 - 20x$
3) $30x^2 - 42x$
4) $21x^2 + 15x$
5) $20x^2 - 15x$
6) $48x^2 + 12x$
7) $3x^2 - 14x + 8$
8) $2x^2 - 4x - 30$
9) $x^2 - 10x + 21$
10) $9x^2 - 16$
11) $25x^2 - 30x + 8$
12) $-8x^4 + 12x^2$
13) $5x^3 - 2x^2 + 3x + 7$
14) $-10x^3 + 4x^2 + 7x$
15) $5x^5 - 6x^3 + 12x^2$
16) $6x^8 + 4x^6 - 5x^2$
17) $-4x^6 + 10x^5 - 12x^3 + 4x$
18) $-16x^3 - 3x^2 + 22$
19) $4x^3 + 2x^2 - 4x$
20) $-x^5 - 4x^2 + 18$
21) $x^6 - 14x^3$
22) $12x^4 - 4$
23) $-9x^4 - 12x^3$
24) -90

25) $-9x^4 - 2x + 19$

26) $-7x^5 - 2x^3 + 17$

27) $11x^3 - 10x + 2$

28) $-8x^4 + 2x^2$

29) $2x^5 + 6x + 2$

30) $11x^5 - 5x^4 + 4$

31) $-9x^5 - 28$

32) $9x^3 - 6x$

Adding and Subtracting Polynomials

1) $x^2 + 1$

2) $6x^3 - 1$

3) $2x^5 + 5x^2 - 15$

4) $6x^3 + 3x^2 - 4x$

5) $-24x^4 + 28x - 6$

6) $14x^2$

7) $14x^2 - 6$

8) $6x^2 - 4$

9) $-3x^3 + 2x$

10) $3x$

11) $19x^3 - 6x$

12) $-7x^5 - 4x$

13) $-11x^7 + 2x^2 + 6x$

14) $4x^5 + 11x^2 - 5$

15) $4x^5 + 18x^4$

16) $-9x^3 + 7x$

17) $-151x^2 + 3x$

18) $-11x^4 + 4x^2 + 5x$

19) $10x^4 - 3x^2 - 4$

20) $10x^6 + 6x^3$

21) $28x^5 + 17x^4 + 22x^2$

22) $10x^4 - 12x$

23) $32x^4 + 4x$

24) $5x^8 - 15x^6 - 5x^3 + 3x$

25) $6x^4 + 15x^2 - 12x$

26) $-2x^9 + 13x^7 + 40x^3$

Multiplying Monomials

1) $-6u^{10}$

2) $10p^{11}$

3) $12xy^3z^9$

4) $24u^6t^5$

5) $35a^5b^6$

6) $-18a^6b^4$

7) $13x^5y^9$

8) $-48p^5q^9$

9) $32s^5t^6$

10) $-36x^6y^4$

11) $36xy^7z^4$

12) $24px^3y^2$

13) $-39p^3q^5$

14) $13s^4t^8$

15) $-66p^8$

16) $-24p^4q^9r^7$

17) $28a^7b$

18) $-30u^9v^6$

19) $-27u^6$

20) $-24x^3y^6$

21) $-13y^8z^4$

22) $16a^5b^2c^6$

23) $35p^9q^8$

24) $-16u^{12}v^6$

25) $-17y^{10}z^6$

26) $-40p^3q^7r^3$

27) $15a^5b^6c^8$

28) $18x^5y^8z^5$

Multiplying and Dividing Monomials

1) $10x^7$

2) $24x^6$

3) $21x^8$

4) $20x^8$

5) $36x^{10}$

6) $32x^{11}y^5$

7) $14x^7y^6$

8) $-10x^6y^9$

9) $18x^7y^7$

10) $-25x^5y^4$

11) $24x^7y^7$

12) $20x^5y^6$

13) $48x^7y^{16}$
14) $45x^7y^{11}$
15) $56x^8y^{14}$
16) $-21x^{12}y^{12}$
17) $5x^5y^2$
18) xy^4
19) $7x^3y^3$
20) $9xy$
21) $4x^5y^3$
22) $12x^5y$
23) $8x^{10}y^3$
24) $5x^{-1}y^3$
25) $5x^3$
26) $-7x^{14}y^5$
27) $-4x^2$

Multiplying a Polynomial and a Monomial

1) $2x^2 + 4x$
2) $-12x + 24$
3) $20x^2 + 10x$
4) $-4x^2 + 5x$
5) $16x^2 - 16x$
6) $12x - 24y$
7) $35x^2 - 35x$
8) $36x^2 + 6xy$
9) $4x^2 + 24xy$
10) $33x^2 + 44xy$
11) $21x^2 + 14x$
12) $40x^2 - 100xy$
13) $27x^2 - 18xy$
14) $7x^2 - 28xy + 42x$
15) $16x^3 + 40xy^2$
16) $24x^2 + 36xy$
17) $8x^4 - 16y^4$
18) $-12x^3y + 16xy$
19) $-20x^3 + 8xy - 16$
20) $4x^2 - 20xy - 24$
21) $16x^4 - 40x^2y + 16x^2$
22) $12x^4 + 36x^2 - 12x^2y$
23) $6x^2 + 3xy - 27y^2$
24) $20x^4 - 12x^2 + 28x$
25) $18x^{22} - 12x - 30$
26) $-2x^5 + 4x^3 + 3x^2$
27) $4x^5 - 2x^3 + 10x^2$
28) $12x^7 - 8x^5 + 20x^4$
29) $8x^6 - 10x^3y + 14x^2y^3$
30) $25x^6 - 15x^3 + 45x^2$
31) $42x^4 + 21x^3 - 42x^2$
32) $4x^4 - 16x^2y + 8xy^2$

Multiplying Binomials

1) $x^2 + 9x + 18$
2) $x^2 - x - 12$
3) $x^2 - 11x + 24$
4) $x^2 + 17x + 72$
5) $x^2 - 14x + 24$
6) $x^2 + 10x + 25$
7) $x^2 + x - 42$
8) $x^2 - 11x + 24$
9) $x^2 + 19x + 84$
10) $x^2 + 4x - 32$
11) $x^2 + 16x + 64$
12) $x^2 + 9x + 14$
13) $x^2 - 36$
14) $x^2 - 25$

15) $x^2 + 22x + 121$
16) $x^2 + 15x + 54$
17) $x^2 - 4$
18) $x^2 + 3x - 28$
19) $3x^2 + 23x + 30$
20) $20x^2 + 16x - 48$
21) $3x^2 - 14x - 49$
22) $x^2 - 13x + 36$
23) $x^2 - 10\text{x} - 24$
24) $10x^2 - 12x - 16$
25) $3x^2 + 16x - 64$
26) $42x^2 + 9x - 6$
27) $12x^2 + 35x + 25$
28) $63x^2 - 8x - 16$
29) $2x^2 - 4x - 16$
30) $25x^2 - 16$
31) $9x^2 - 15x - 14$
32) $x^4 - 64$

Factoring Trinomials

1) $(x + 6)(x + 2)$
2) $(x - 5)(x - 1)$
3) $(x + 12)(x + 3)$
4) $(x - 5)(x - 7)$
5) $(x - 2)(x - 9)$
6) $(x - 6)(x - 3)$
7) $(x + 6)(x + 12)$
8) $(x + 8)(x - 9)$
9) $(x - 3)(x + 7)$
10) $(x - 11)(x - 2)$
11) $(x - 4)(x + 6)$
12) $(x - 8)(x + 5)$
13) $(x + 7)(x - 10)$
14) $(x + 13)(x + 13)$
15) $(4x + 5)(x - 3)$
16) $(x - 11)(x - 3)$
17) $(5x - 5)(2x + 3)$
18) $(2x - 6)(3x + 7)$
19) $(x + 6)(x + 6)$
20) $(5x - 3)(x + 4)$
21) $(x - 8)$
22) $(4x - 3)$
23) $(2x - 6)$

Operations with Polynomials

1) $20x + 12$
2) $16x + 48$
3) $10x - 4$
4) $-28x + 12$
5) $27x^3 + 3x^2$
6) $28x^7 - 36x^6$
7) $-21x^5 + 9x^4$
8) $-40x^5 + 64x^4$
9) $7x^2 + 35x - 21$
10) $45x^2 - 63x + 45$
11) $9x^2 + 9x + 6$
12) $15x^3 + 25x^2 + 40x$
13) $15x^2 + 6x - 21$
14) $27x^2 - 36x - 15$
15) $24x^2 - 6$
16) $21x^2 + 29x - 10$
17) $(5x + 2y)$
18) $16x^2 - 2x - \frac{15}{2}$
19) $36x^2 + 24x + 4$
20) $40x$
21) $x^3 + 6x^2 + 12x + 8$
22) $(9x + y)$

Chapter 9 :

Complex Numbers

Topics that you'll practice in this chapter:

- ✓ Adding and Subtracting Complex Numbers
- ✓ Multiplying and Dividing Complex Numbers
- ✓ Graphing Complex Numbers
- ✓ Rationalizing Imaginary Denominators

Mathematics is a hard thing to love. It has the unfortunate habit, like a rude dog, of turning its most unfavorable side towards you when you first make contact with it. — David Whiteland

Adding and Subtracting Complex Numbers

Simplify.

1) $(7i) - (3i) =$

2) $(5i) + (4i) =$

3) $(2i) + (8i) =$

4) $(-8i) - (3i) =$

5) $(14i) + (6i) =$

6) $(6i) - (-10i) =$

7) $(-2i) + (-5i) =$

8) $(13i) - (5i) =$

9) $(-22i) - (11i) =$

10) $(-4i) + (2 + 6i) =$

11) $(10 - 5i) + (-3i) =$

12) $(-8i) + (6 + 12i) =$

13) $1 + (5 - 4i) =$

14) $(13i) - (-8 + 2i) =$

15) $(8 + 12i) - (-10i) =$

16) $(10 + i) + (-5i) =$

17) $(11i) - (-7i + 9) =$

18) $(10i + 12) + (-2i) =$

19) $(20) - (16 + 4i) =$

20) $(3 + 3i) + (8 + 4i) =$

21) $(15 - 7i) + (3 + 4i) =$

22) $(12 + 6i) + (10 + 17i) =$

23) $(-5 + 6i) - (-16 - 12i) =$

24) $(-4 + 14i) - (-9 + 11i) =$

25) $(-22 + 4i) - (-7 - 22i) =$

26) $(-26 - 18i) + (3 + 34i) =$

27) $(-19 - 13i) - (-7 - 20i) =$

28) $-21 + (5i) + (-32 + 14i) =$

29) $30 - (7i) + (3 - 11i) =$

30) $28 + (-32 - 10i) - 7 =$

31) $(-44i) + (2 - 7i) + 9 =$

32) $(-21i) - (12 - 9i) + 21i =$

Multiplying and Dividing Complex Numbers

Simplify.

1) $(5i)(-3i) =$

2) $(-8i)(2i) =$

3) $(3i)(-3i)(-3i) =$

4) $(6i)(-6i) =$

5) $(-3-4i)(2+i) =$

6) $(5-2i)^2 =$

7) $(5-2i)(6-4i) =$

8) $(1+6i)^2 =$

9) $(5i)(-3i)(2-4i) =$

10) $(11-2i)(2-4i) =$

11) $(-3+i)(6+5i) =$

12) $(2-8i)(6-4i) =$

13) $3(4i)-(6i)(-2+5i) =$

14) $\frac{5}{-25i} =$

15) $\frac{3-4i}{-5i} =$

16) $\frac{6+12i}{2i} =$

17) $\frac{20i}{-5+4i} =$

18) $\frac{-6-9i}{4i} =$

19) $\frac{4i}{8-2i} =$

20) $\frac{4-7i}{6-2i} =$

21) $\frac{3-2i}{-1-1i} =$

22) $\frac{-5-5i}{-4-i} =$

23) $\frac{-6+2i}{-10-4i} =$

24) $\frac{-8-4i}{-2+4i} =$

25) $\frac{2+3i}{1-4i} =$

Graphing Complex Numbers

Identify each complex number graphed.

1)

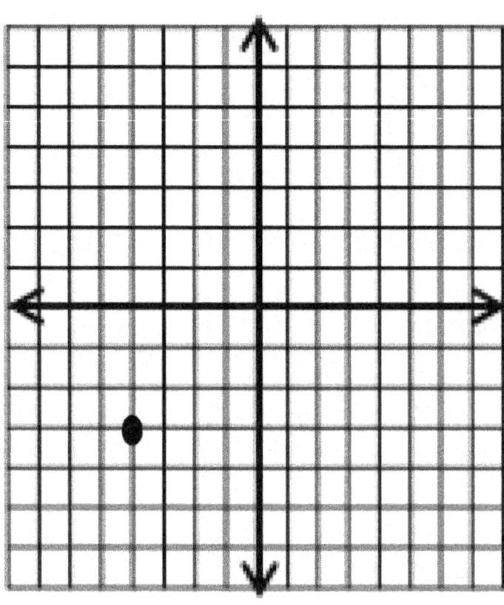

2)

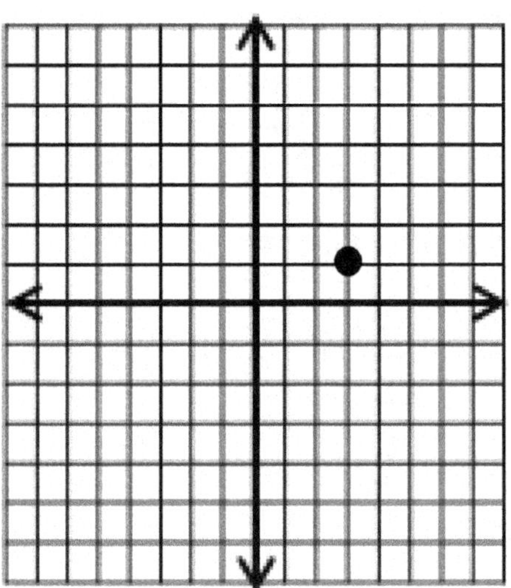

3)

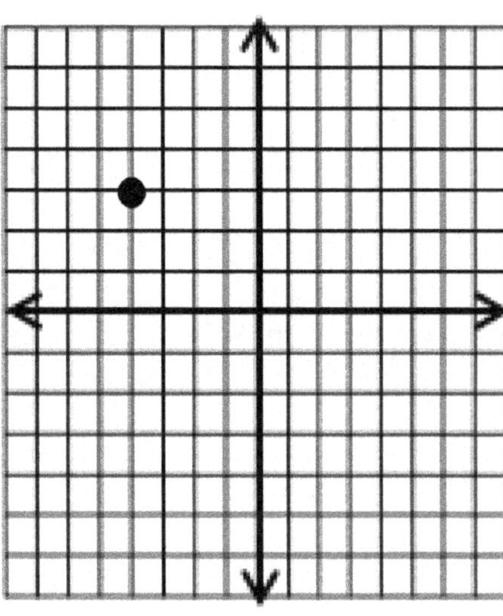

4)

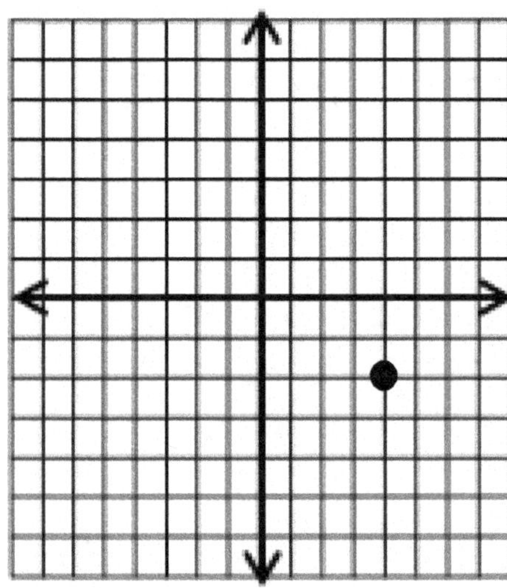

Rationalizing Imaginary Denominators

Simplify.

1) $\frac{-7}{-7i} =$

2) $\frac{-3}{-15i} =$

3) $\frac{-3}{-39i} =$

4) $\frac{24}{-3i} =$

5) $\frac{5}{2i} =$

6) $\frac{16}{-4i} =$

7) $\frac{14}{-6i} =$

8) $\frac{-17}{3i} =$

9) $\frac{4x}{5yi} =$

10) $\frac{10-10i}{-2i} =$

11) $\frac{5-11i}{-i} =$

12) $\frac{21+4i}{4i} =$

13) $\frac{8i}{-1+4i} =$

14) $\frac{10i}{-6+8i} =$

15) $\frac{-25-5i}{-5+5i} =$

16) $\frac{-7-2i}{3+1i} =$

17) $\frac{-12-6i}{8-6i} =$

18) $\frac{-14+7i}{-7i} =$

19) $\frac{12+3i}{3i} =$

20) $\frac{-2-i}{4-3i} =$

21) $\frac{-11+4i}{-5i} =$

22) $\frac{8+2i}{-5-2i} =$

23) $\frac{-9-5i}{-8-2i} =$

24) $\frac{4i-1}{-5-2i} =$

Answers of Worksheets – Chapter 9

Adding and Subtracting Complex Numbers

1) $4i$
2) $9i$
3) $10i$
4) $-11i$
5) $20i$
6) $16i$
7) $-7i$
8) $8i$
9) $-33i$
10) $2+2i$
11) $10-8i$
12) $6+4i$
13) $6-4i$
14) $8+11i$
15) $8+22i$
16) $10-4i$
17) $-9+18i$
18) $12+8i$
19) $4-4i$
20) $11+7i$
21) $18-3i$
22) $22+23i$
23) $11+18i$
24) $5+3i$
25) $-15+26i$
26) $-23+16i$
27) $-12+7i$
28) $-53+19i$
29) $33-18i$
30) $-11-10i$
31) $11-51i$
32) $-12+9i$

Multiplying and Dividing Complex Numbers

1) 15
2) 16
3) $-27i$
4) 36
5) $-2-11i$
6) $21-20i$
7) $22-32i$
8) $-35+12i$
9) $30-60i$
10) $14-48i$
11) $-23-9i$
12) $-20-56i$
13) $30+24i$
14) $\frac{i}{5}$
15) $\frac{4}{5}+\frac{3}{5}i$
16) $-6+3i$
17) $\frac{80}{41}-\frac{100}{41}i$
18) $\frac{9}{4}-\frac{3}{2}i$
19) $-\frac{2}{17}+\frac{8}{17}i$
20) $\frac{19}{20}-\frac{17}{20}i$
21) $-\frac{1}{2}+\frac{5}{2}i$
22) $\frac{25}{17}+\frac{15}{17}i$
23) $\frac{13}{29}-\frac{11}{29}i$
24) $2i$
25) $-\frac{10}{17}+\frac{11}{17}i$

Graphing Complex Numbers

1) $-4-3i$
2) $3+i$
3) $-4+3i$
4) $4-2i$

Rationalizing Imaginary Denominators

1) $-i$
2) $-\frac{1}{5}i$
3) $\frac{-1}{13}i$
4) $8i$
5) $-\frac{5}{2}i$
6) $4i$
7) $\frac{7}{3}i$
8) $\frac{17}{3}i$
9) $-\frac{4x}{5y}i$
10) $5+5i$
11) $11+5i$
12) $1-\frac{21}{4}i$
13) $\frac{32}{17}-\frac{8}{17}i$
14) $\frac{4}{5}-\frac{3}{5}i$
15) $2+3i$
16) $-\frac{23}{10}+\frac{1}{10}i$
17) $-\frac{3}{5}-\frac{6}{5}i$
18) $-1-2i$
19) $1-4i$
20) $-\frac{1}{5}-\frac{2}{5}i$
21) $-\frac{4}{5}-\frac{11}{5}i$
22) $-\frac{44}{29}+\frac{6}{29}i$
23) $\frac{41}{34}+\frac{11}{34}i$
24) $-\frac{3}{29}-\frac{22}{29}i$

Chapter 10 :

Functions Operations and Quadratic

Topics that you'll practice in this chapter:

- ✓ Evaluating Function
- ✓ Adding and Subtracting Functions
- ✓ Multiplying and Dividing Functions
- ✓ Composition of Functions
- ✓ Quadratic Equation
- ✓ Solving Quadratic Equations
- ✓ Quadratic Formula and the Discriminant
- ✓ Quadratic Inequalities
- ✓ Graphing Quadratic Functions
- ✓ Domain and Range of Radical Functions
- ✓ Solving Radical Equations

It's fine to work on any problem, so long as it generates interesting mathematics along the way – even if you don't solve it at the end of the day." – Andrew Wiles

Evaluating Function

Write each of following in function notation.

1) $h = -8x + 3$

2) $k = 2a - 14$

3) $d = 11t$

4) $y = \frac{5}{12}x - \frac{7}{12}$

5) $m = 24n - 210$

6) $c = p^2 - 5p + 10$

Evaluate each function.

7) $f(x) = 2x - 7$, find $f(-3)$

8) $g(x) = \frac{1}{9}x + 12$, find $f(18)$

9) $h(x) = -4x + 9$, find $f(3)$

10) $f(x) = -x + 19$, find $f(-3)$

11) $f(a) = 7a - 12$, find $f(3)$

12) $h(x) = 14 - 3x$, find $f(-4)$

13) $g(n) = 6n - 10$, find $f(2)$

14) $f(x) = -11x - 4$, find $f(-1)$

15) $k(n) = -20 - 3.5n$, find $f(2)$

16) $f(x) = -0.7x + 3.3$, find $f(-7)$

17) $g(n) = \frac{11n+8}{n}$, find $g(2)$

18) $g(n) = \sqrt{3n} + 12$, find $g(3)$

19) $h(x) = x^{-2} - 7$, find $h(\frac{1}{9})$

20) $h(n) = n^{-3} + 11$, find $h(\frac{1}{4})$

21) $h(n) = n^3 - 2$, find $h(\frac{1}{2})$

22) $h(n) = n^2 - 4$, find $h(-\frac{1}{3})$

23) $h(n) = 4n^2 - 13$, find $h(-5)$

24) $h(n) = -2n^3 - 6n$, find $h(2)$

25) $g(n) = \sqrt{16n^2} - \sqrt{n}$, find $g(4)$

26) $h(a) = \frac{-14a+9}{3a}$, find $h(-b)$

27) $k(a) = 12a - 14$, find $k(a-3)$

28) $h(x) = \frac{1}{9}x + 18$, find $h(-18x)$

29) $h(x) = 8x^2 + 16$, find $h(\frac{x}{2})$

30) $h(x) = x^4 - 20$, find $h(-2x)$

Adding and Subtracting Functions

Perform the indicated operation.

1) $f(x) = 2x + 3$

$g(x) = x + 7$

Find $(f - g)(2)$

2) $g(a) = -5a - 8$

$f(a) = -3a - 5$

Find $(g - f)(-2)$

3) $h(t) = 4t + 3$

$g(t) = 4t + 7$

Find $(h - g)(t)$

4) $g(a) = -6a - 10$

$f(a) = 3a^2 + 9$

Find $(g - f)(x)$

5) $g(x) = \frac{5}{6}x - 23$

$h(x) = \frac{5}{12}x + 25$

Find $g(12) - h(12)$

6) $h(x) = \sqrt{3x} - 2$

$g(x) = \sqrt{3x} + 5$

Find $(h + g)(12)$

7) $f(x) = x^{-1}$

$g(x) = x^2 + \frac{5}{x}$

Find $(f - g)(-3)$

8) $h(n) = n^2 + 2$

$g(n) = -4n + 6$

Find $(h - g)(2a)$

9) $g(x) = -2x^2 - 5 - 4x$

$f(x) = 7 + 2x$

Find $(g - f)(3x)$

10) $g(t) = 11t - 4$

$f(t) = -2t^2 + 5$

Find $(g + f)(-t)$

11) $f(x) = 8x + 9$

$g(x) = -5x^2 + 3x$

Find $(f - g)(-x^2)$

12) $f(x) = -3x^4 - 5x$

$g(x) = 2x^4 + 5x + 22$

Find $(f + g)(3x^2)$

Multiplying and Dividing Functions

Perform the indicated operation.

1) $g(x) = -2x - 1$

$f(x) = 4x + 3$

Find $(g.f)(2)$

2) $f(x) = 5x$

$h(x) = -2x + 3$

Find $(f.h)(-2)$

3) $g(a) = 5a - 2$

$h(a) = 2a - 3$

Find $(g.h)(-3)$

4) $f(x) = 2x - 7$

$h(x) = x - 5$

Find $(\frac{f}{h})(4)$

5) $f(x) = 8a^2$

$g(x) = 3 + 2a$

Find $(\frac{f}{g})(2)$

6) $g(a) = \sqrt{4a} + 2$

$f(a) = (-a)^4 + 1$

Find $(\frac{g}{f})(1)$

7) $g(t) = t^3 + 1$

$h(t) = 5t - 2$

Find $(g.h)(-2)$

8) $g(n) = n^2 + 2n - 4$

$h(n) = -5n + 3$

Find $(g.h)(1)$

9) $g(a) = (a - 3)^2$

$f(a) = a^2 + 4$

Find $(\frac{g}{f})(3)$

10) $g(x) = -3x^2 + \frac{4}{5}x + 9$

$f(x) = x^2 - 24$

Find $(\frac{g}{f})(5)$

11) $f(x) = 2x^3 - 5x^2 + 1$

$g(x) = 3x - 1$

Find $(f.g)(x)$

12) $f(x) = 5x - 2$

$g(x) = x^3 - 2x$

Find $(f.g)(x^2)$

Composition of Functions

Using $f(x) = 2x - 5$ and $g(x) = -2x$, find:

1) $f(g(2)) =$

2) $f(g(-1)) =$

3) $g(f(-4)) =$

4) $g(f(5)) =$

5) $f(g(3)) =$

6) $g(f(0)) =$

Using $f(x) = -\frac{1}{4}x + \frac{3}{4}$ and $g(x) = 2x^2$, find:

7) $g(f(-2)) =$

8) $g(f(4)) =$

9) $g(g(1)) =$

10) $f(f(1)) =$

11) $g(f(-4)) =$

12) $f(g(x)) =$

Using $f(x) = -2x + 2$ and $g(x) = x + 1$, find:

13) $g(f(1)) =$

14) $f(f(0)) =$

15) $f(g(-1)) =$

16) $f(g(-3)) =$

17) $g(f(2)) =$

18) $f(g(x)) =$

Using $f(x) = \sqrt{x + 9}$ and $g(x) = x - 9$, find:

19) $f(g(9)) =$

20) $g(f(-9)) =$

21) $f(g(4)) =$

22) $f(f(7)) =$

23) $g(f(-5)) =$

24) $g(g(0)) =$

Quadratic Equation

Multiply.

1) $(x-4)(x+6) =$ ____________

2) $(x+5)(x+7) =$ ____________

3) $(x-6)(x+8) =$ ____________

4) $(x+2)(x-9) =$ ____________

5) $(x-7)(x-8) =$ ____________

6) $(3x+2)(x-3) =$ ____________

7) $(4x-3)(x+2) =$ ____________

8) $(4x-5)(x+1) =$ ____________

9) $(7x+1)(x-6) =$ ____________

10) $(5x+1)(3x-3) =$ ________

Factor each expression.

11) $x^2-2x-8 =$ ____________

12) $x^2+8x+15 =$ ____________

13) $x^2-2x-24 =$ ____________

14) $x^2-10x+21 =$ ____________

15) $x^2+10x+21 =$ ____________

16) $4x^2+9x+5 =$ ____________

17) $5x^2+13x-6 =$ ____________

18) $5x^2+17x-12 =$ ____________

19) $2x^2+7x+5 =$ ____________

20) $9x^2-21x+6 =$ ____________

Calculate each equation.

21) $(x+6)(x-3) = 0$

22) $(x+1)(x+8) = 0$

23) $(3x+6)(x+5) = 0$

24) $(2x-2)(4x+8) = 0$

25) $x^2+x+10 = 22$

26) $x^2+11x+36 = 12$

27) $2x^2+9x+9 = 5$

28) $x^2+3x-24 = 4$

29) $5x^2+5x-40 = 20$

30) $8x^2+8x = 48$

Solving Quadratic Equations

Solve each equation by factoring or using the quadratic formula.

1) $(x+9)(x-1)=0$

2) $(x+7)(x+6)=0$

3) $(x-8)(x+3)=0$

4) $(x-6)(x-4)=0$

5) $(x+2)(x+12)=0$

6) $(5x+4)(x+7)=0$

7) $(6x+1)(4x+5)=0$

8) $(2x+7)(x+8)=0$

9) $(x+6)(3x+15)=0$

10) $(12x+2)(x+8)=0$

11) $x^2=8x$

12) $x^2-16=0$

13) $3x^2+6=9x$

14) $-2x^2-8=10x$

15) $5x^2+40x=45$

16) $x^2+10x=24$

17) $x^2+6x=16$

18) $x^2+9x=-18$

19) $x^2+13x=-36$

20) $x^2+3x-15=5x$

21) $x^2+8x+7=-8$

22) $3x^2-11x=-9+x$

23) $10x^2+3=27x-15$

24) $7x^2-6x+8=8$

25) $2x^2-12=-3x+2$

26) $10x^2-26x-3=-15$

27) $3x^2+21=-16x+5$

28) $x^2+15x-10=-66$

29) $3x^2-8x-8=4+x$

30) $2x^2+6x-24=12$

31) $3x^2-33x+54=-18$

32) $-10x^2-15x-9=-9-27x^2$

Quadratic Formula and the Discriminant

Find the value of the discriminant of each quadratic equation.

1) $3x(x-8)=0$

2) $2x^2+6x-4=0$

3) $x^2+6x+7=0$

4) $x^2-x+3=0$

5) $x^2+4x-3=0$

6) $2x^2+6x-10=0$

7) $3x^2+7x+5=0$

8) $x^2-6x-4=0$

9) $2x^2+8x+3=0$

10) $x^2+7x-5=0$

11) $5x^2+2x-3=0$

12) $-3x^2-11x+4=0$

13) $-6x^2-12x+8=0$

14) $-x^2-9x-12=0$

15) $7x^2-6x-10=0$

16) $-4x^2-2x+8=0$

17) $5x^2+8x-2=0$

18) $6x^2-4x=0$

19) $3x^2-5x+2=0$

20) $4x^2+9x+3=0$

Find the discriminant of each quadratic equation then state the number of real and imaginary solutions.

21) $-4x^2-16=16x$

22) $20x^2=20x-5$

23) $-11x^2-19x=26$

24) $22x^2-4x+1=18x^2$

25) $-11x^2=-15x+8$

26) $3x^2+6x+9=6$

27) $13x^2-5x-12=-26$

28) $-8x^2-32x-25=7$

Graphing Quadratic Functions

Sketch the graph of each function. Identify the vertex and axis of symmetry.

1) $y = (x+3)^2 + 2$

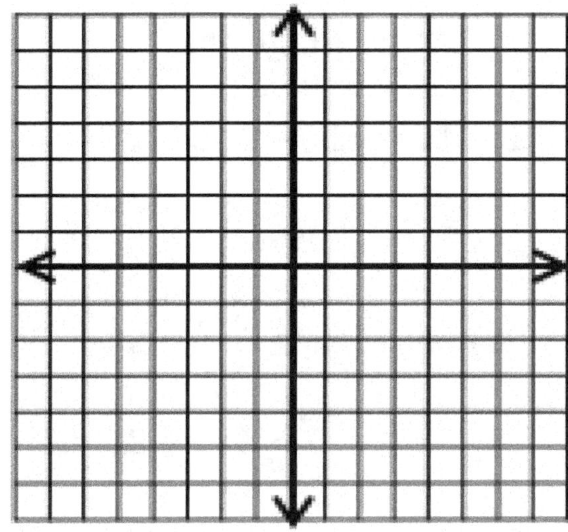

2) $y = (x-3)^2 - 2$

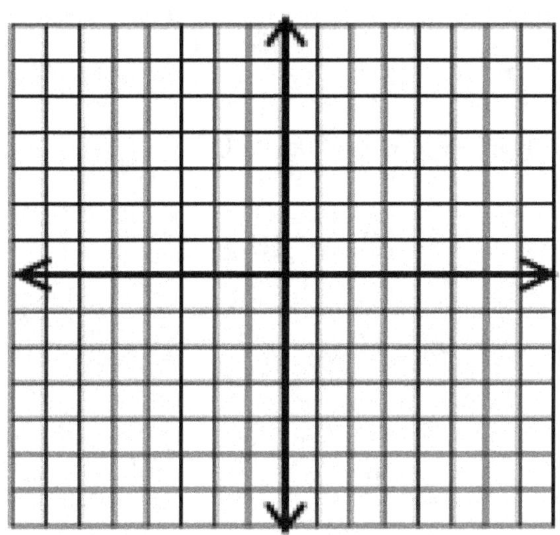

3) $y = 6 - (-x+4)^2$

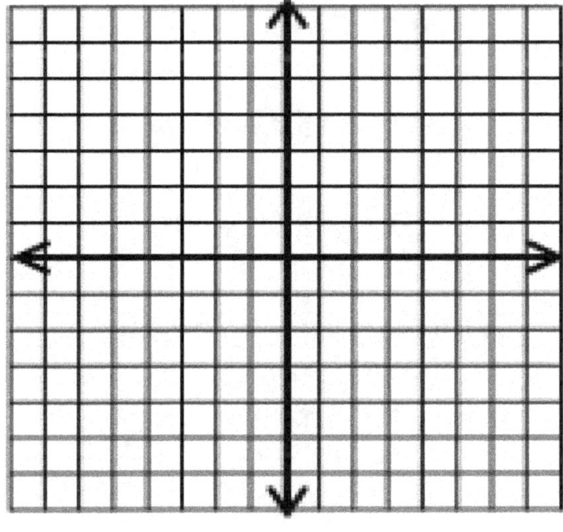

4) $y = -3x^2 - 6x + 9$

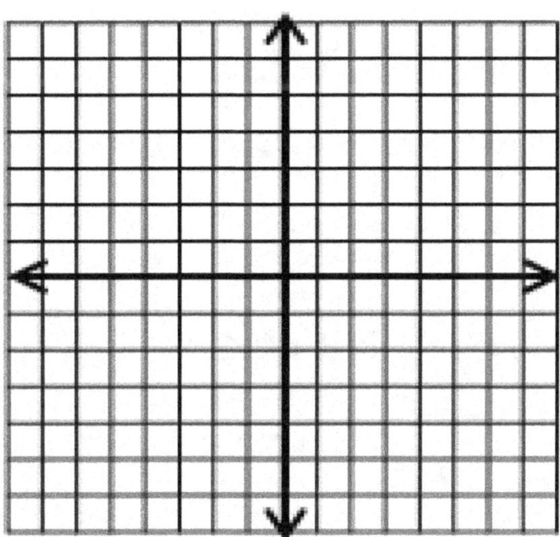

Quadratic Inequalities

Solve each quadratic inequality.

1) $x^2 - 25 < 0$

2) $-x^2 - 6x - 8 > 0$

3) $5x^2 + 15x + 30 < 0$

4) $x^2 + 8x + 16 > 0$

5) $2x^2 - 18x - 20 \geq 0$

6) $x^2 > -10x - 25$

7) $3x^2 + 2x + 16 \leq 0$

8) $x^2 - 5x - 14 \leq 0$

9) $x^2 - 6x - 7 \geq 0$

10) $2x^2 + 16x - 18 < 0$

11) $x^2 + 6x - 72 > 0$

12) $3x^2 - 3x - 36 > 0$

13) $x^2 - 15x + 64 \leq 0$

14) $2x^2 - 24x + 72 \leq 0$

15) $x^2 - 16x + 63 \geq 0$

16) $x^2 - 16x + 55 \geq 0$

17) $x^2 - 81 \leq 0$

18) $x^2 - 17x + 42 \geq 0$

19) $9x^2 + 14x + 36 \leq 0$

20) $4x^2 - 2x - 24 > 2x^2$

21) $5x^2 - 20x + 20 < 0$

22) $7x^2 - 6x \geq 6x^2 - 5$

23) $5x^2 - 15 > 4x^2 + 2x$

24) $3x^2 - 4x \geq 3x^2 - 9x + 15$

25) $8x^2 + 9x - 54 > 5x^2$

26) $10x^2 + 50x - 60 < 0$

27) $-x^2 + 15x - 57 \geq 0$

28) $-5x^2 + 25x + 30 \leq 0$

29) $5x^2 + 40x + 75 < 0$

30) $9x^2 + 20x + 180 \leq 0$

31) $3x^2 + 2x - 36 \geq -x$

32) $3x^2 + 9x + 9 \leq 6x^2 + 3x$

Domain and Range of Radical Functions

Identify the domain and range of each function.

1) $y = \sqrt{x+8} - 7$

2) $y = \sqrt[3]{3x-5} - 4$

3) $y = \sqrt{3x-9} + 3$

4) $y = \sqrt[3]{(4x+6)} - 2$

5) $y = 3\sqrt{4x+20} + 6$

6) $y = \sqrt[3]{(5x-2)} - 11$

7) $y = 4\sqrt{9x^2+8} + 3$

8) $y = \sqrt[3]{(7x^2-2)} - 6$

9) $y = 2\sqrt{2x^3+16} - 3$

10) $y = \sqrt[3]{(11x+4)} - 2x$

11) $y = 3\sqrt{-2(4x+8)} + 5$

12) $y = \sqrt[5]{(3x^2-12)} - 6$

13) $y = 3\sqrt{x-5} - 2$

14) $y = \sqrt[3]{6x+9} - 4$

Sketch the graph of each function.

15) $y = -3\sqrt{x} + 5$

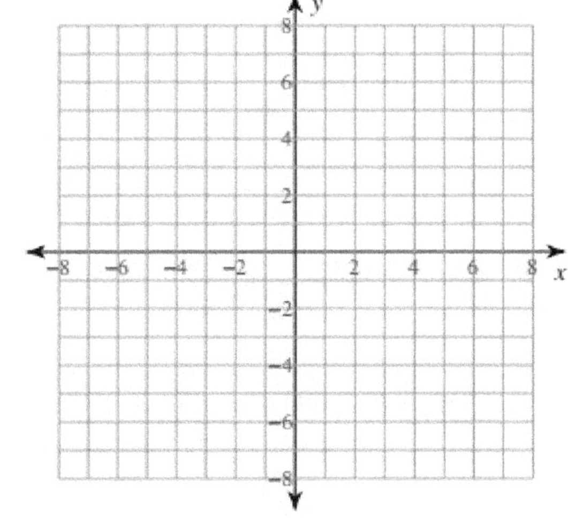

16) $y = 3\sqrt{x} - 6$

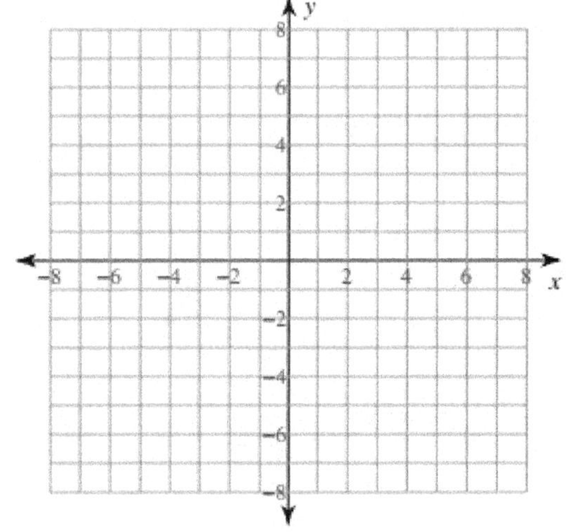

Solving Radical Equations

Solve each equation. Remember to check for extraneous solutions.

1) $\sqrt{a} = 9$

2) $\sqrt{v} = 6$

3) $\sqrt{r} = 4$

4) $8 = 16\sqrt{x}$

5) $\sqrt{x+3} = 18$

6) $6 = \sqrt{x-7}$

7) $4 = \sqrt{r-3}$

8) $\sqrt{x-5} = 7$

9) $12 = \sqrt{x-4}$

10) $\sqrt{m+5} = 8$

11) $7\sqrt{5a} = 35$

12) $6\sqrt{2x} = 48$

13) $2 = \sqrt{6x-32}$

14) $\sqrt{304-4x} = 4$

15) $\sqrt{r+2} - 8 = 6$

16) $-21 = -7\sqrt{r+9}$

17) $60 = 6\sqrt{5v}$

18) $x = \sqrt{40-3x}$

19) $\sqrt{90-27a} = 3a$

20) $\sqrt{-8n+88} = 4$

21) $\sqrt{15r-5} = 4r-3$

22) $\sqrt{-64+32x} = 4x$

23) $\sqrt{4x+15} = \sqrt{2x+11}$

24) $\sqrt{12v} = \sqrt{15v-21}$

25) $\sqrt{9-x} = \sqrt{x-3}$

26) $\sqrt{6m+34} = \sqrt{8m+34}$

27) $\sqrt{7r+32} = \sqrt{-8-3r}$

28) $\sqrt{4k+10} = \sqrt{2-4k}$

29) $-20\sqrt{x-13} = -40$

30) $\sqrt{90-2x} = \sqrt{\frac{x}{4}}$

Answers of Worksheets

Evaluating Function

1) $h(x) = -8x + 3$
2) $k(a) = 2a - 14$
3) $d(t) = 11t$
4) $f(x) = \frac{5}{12}x - \frac{7}{12}$
5) $m(n) = 24n - 210$
6) $c(p) = p^2 - 5p + 10$
7) -13
8) 14
9) -3
10) 22
11) 9
12) 26
13) 2
14) 7
15) -27
16) 8.2
17) 15
18) 15
19) 74
20) 75
21) $-1\frac{7}{8}$
22) $-3\frac{8}{9}$
23) 87
24) -28
25) 14
26) $-\frac{14b+9}{3b}$
27) $12a - 50$
28) $-2x + 18$
29) $2x^2 + 16$
30) $16x^4 - 20$

Adding and Subtracting Functions

1) -2
2) 1
3) -4
4) $-3x^2 - 6x - 19$
5) -43
6) 15
7) $-7\frac{2}{3}$
8) $4a^2 + 8a - 4$
9) $-18x^2 - 18x - 12$
10) $-2t^2 - 11t + 1$
11) $5x^4 - 5x^2 + 9$
12) $-81x^8 + 22$

Multiplying and Dividing Functions

1) -55
2) -70
3) 153
4) -1
5) $4\frac{4}{7}$
6) 2
7) 84
8) 2
9) 0
10) -62
11) $6x^4 - 17x^3 + 5x^2 + 3x - 1$
12) $5x^8 - 2x^6 - 10x^4 + 4x^2$

Composition of Functions

1) -13
2) -1
3) 26
4) -10
5) -17
6) 10
7) $\frac{25}{8}$
8) $\frac{1}{8}$
9) 8
10) $\frac{5}{8}$
11) $\frac{49}{8}$
12) $-\frac{1}{2}(x^2 - \frac{3}{2})$
13) 1
14) -2
15) 2
16) 6
17) -1
18) $-2x$

19) 3
20) -9
21) 2
22) $\sqrt{13}$
23) -7
24) -18

Quadratic Equations

1) $x^2 + 2x - 24$
2) $x^2 + 12x + 35$
3) $x^2 + 2x - 48$
4) $x^2 - 7x - 18$
5) $x^2 - 15x + 56$
6) $3x^2 - 7x - 6$
7) $4x^2 + 5x - 6$
8) $4x^2 - x - 5$
9) $7x^2 - 41x - 6$
10) $15x^2 - 12x - 3$
11) $(x - 4)(x + 2)$
12) $(x + 5)(x + 3)$
13) $(x - 6)(x + 4)$
14) $(x - 3)(x - 7)$
15) $(x + 3)(x + 7)$
16) $(4x + 5)(x + 1)$
17) $(5x - 2)(x + 3)$
18) $(5x - 3)(x + 4)$
19) $(2x + 5)(x + 1)$
20) $3(x - 2)(3x - 1)$
21) $x = -6, x = 3$
22) $x = -1, x = -8$
23) $x = -2, x = -5$
24) $x = 1, x = -2$
25) $x = 3, x = -4$
26) $x = -3, x = -8$
27) $x = -4, x = -\frac{1}{2}$
28) $x = 4, x = -7$
29) $x = 3, x = -4$
30) $x = -3, x = 2$

Solving quadratic equations

1) $\{-9, 1\}$
2) $\{-6, -7\}$
3) $\{8, -3\}$
4) $\{6, 4\}$
5) $\{-2, -12\}$
6) $\{-\frac{4}{5}, -7\}$
7) $\{-\frac{5}{4}, -\frac{1}{6}\}$
8) $\{-\frac{7}{2}, -8\}$
9) $\{-6, -5\}$
10) $\{-\frac{1}{6}, -8\}$
11) $\{8, 0\}$
12) $\{4, -4\}$
13) $\{2, 1\}$
14) $\{-4, -1\}$
15) $\{1, -9\}$
16) $\{2, -12\}$
17) $\{2, -8\}$
18) $\{-3, -6\}$
19) $\{-4, -9\}$
20) $\{5, -3\}$
21) $\{-5, -3\}$
22) $\{1, 3\}$
23) $\{\frac{6}{5}, \frac{3}{2}\}$
24) $\{\frac{6}{7}, 0\}$
25) $\{-\frac{7}{2}, 2\}$
26) $\{\frac{3}{5}, 2\}$
27) $\{-\frac{4}{3}, -4\}$
28) $\{-8, -7\}$
29) $\{4, -1\}$
30) $\{3, -6\}$
31) $\{3, 8\}$
32) $\{\frac{15}{17}, 0\}$

Quadratic formula and the discriminant

1) 576
2) 68
3) 8
4) -11
5) 28
6) 116
7) -11
8) 52
9) 40
10) 69
11) 64
12) 169
13) 336
14) 33
15) 316
16) 132
17) 104
18) 16
19) 1
20) 33
21) 0, *one real solution*
22) 0, *one real solution*
23) -783, *no solution*

24) $0, one\ real\ solution$

25) $-127, no\ solution$

26) $0, one\ real\ solution$

27) $-703, no\ solution$

28) $0, one\ real\ solution$

Graphing quadratic functions

1) $(-3, 2), x = -3$

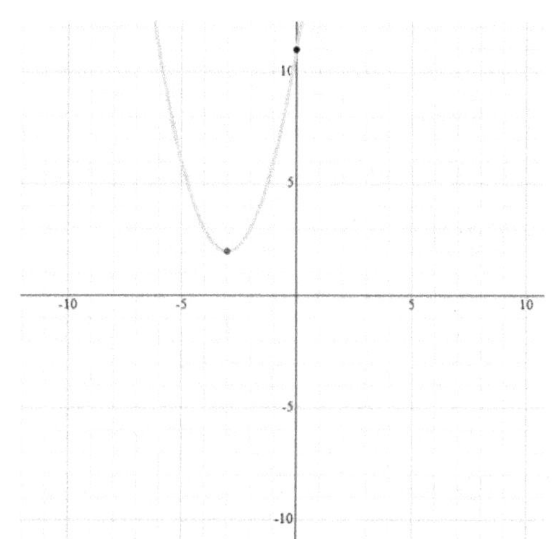

2) $(3, -2), x = 3$

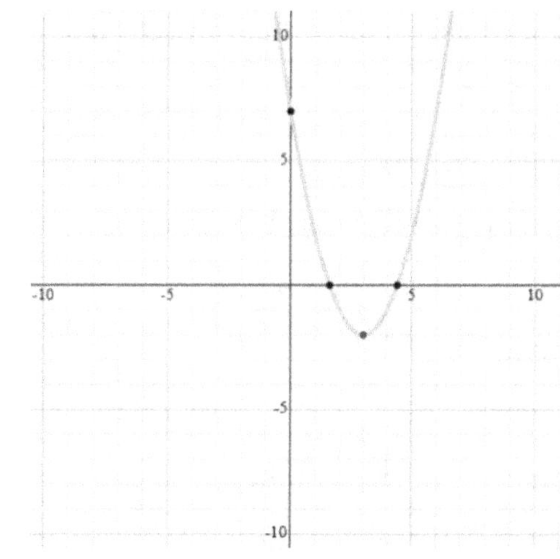

3) $(4, 6), x = 4$

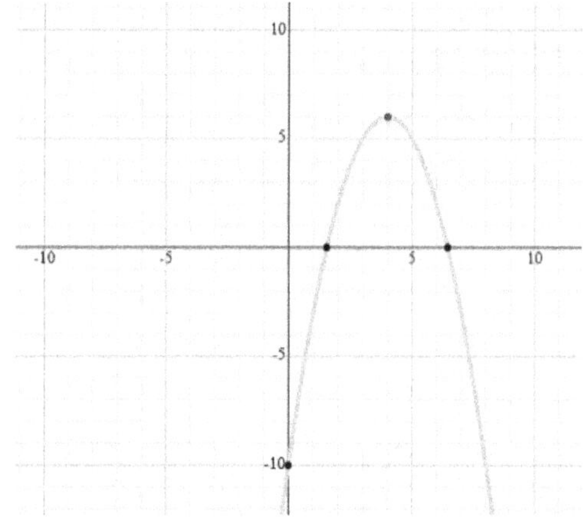

4) $(-1, 12), x = -1$

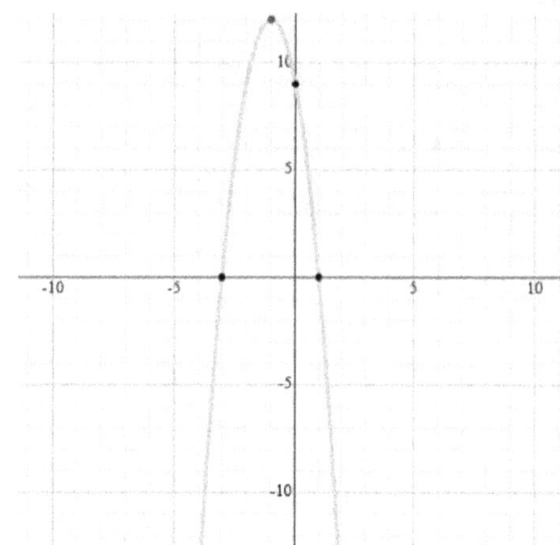

Quadratic inequalities

1) $-5 < x < 5$
2) $-4 < x < -2$
3) no solution
4) $x < -4\ or\ x > -4$
5) $x \leq -1\ or\ x \geq 10$
6) $x < -5\ or\ x > -5$
7) no solution
8) $-2 \leq x \leq 7$
9) $x \leq -1\ or\ x \geq 7$
10) $-9 < x < 1$
11) $x < -12\ or\ x > 6$
12) $-3 < x < 4$
13) no solution
14) $x = 6$
15) $x \leq 7 or\ \ x \geq 9$
16) $x \leq 5 or\ \ x \geq 11$
17) $-9 \leq x \leq 9$
18) $x \leq 3\ or\ x \geq 14$

19) no solution

20) $x < -3$ *or* $x > 4$

21) no solution

22) $x \leq 1$ *or* $x \geq 5$

23) $x < -3$ *or* $x > 5$

24) $x \geq 3$

25) $x < -6$ *or* $x > 3$

26) $-6 < x < 1$

27) no solution

28) $x \leq -1$ or $x \geq 6$

29) $-5 < x < -3$

30) no solution

31) $x \leq -4$ or $x \geq 3$

32) $x \leq -1$ *or* $x \geq 3$

Domain and range of radical functions

1) domain: $x \geq -8$
   range: $y \geq -7$
2) domain: {all real numbers}
   range: {all real numbers}
3) domain: $x \geq 3$
   range: $y \geq 3$
4) domain: {all real numbers}
   range: {all real numbers}
5) domain: $x \geq -5$
   range: $y \geq 6$
6) domain: {all real numbers}
   range: {all real numbers}
7) domain: {all real numbers}
   range: $y \geq 8\sqrt{2} + 3$
8) domain: {all real numbers}
   range: {all real numbers}
9) domain: $x \geq -2$
   range: $y \geq -3$
10) domain: {all real numbers}
    range: {all real numbers}
11) domain: $x \leq -2$
    range: $y \geq 5$
12) domain: {all real numbers}
    range: {all real numbers}
13) domain: $x \geq 5$
    range: $y \geq -2$
14) domain: {all real numbers}
    range: {all real numbers}

15)

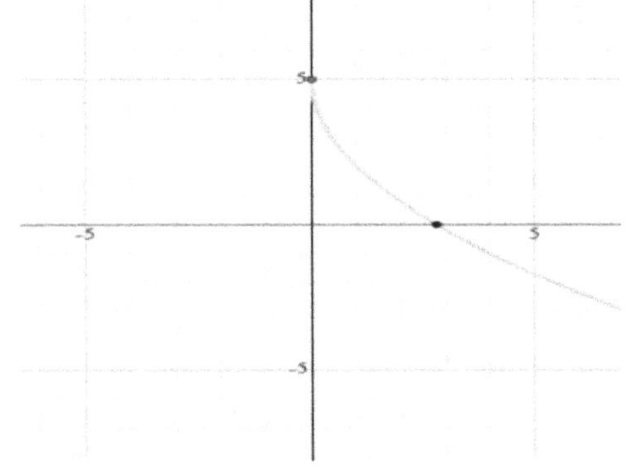

16)

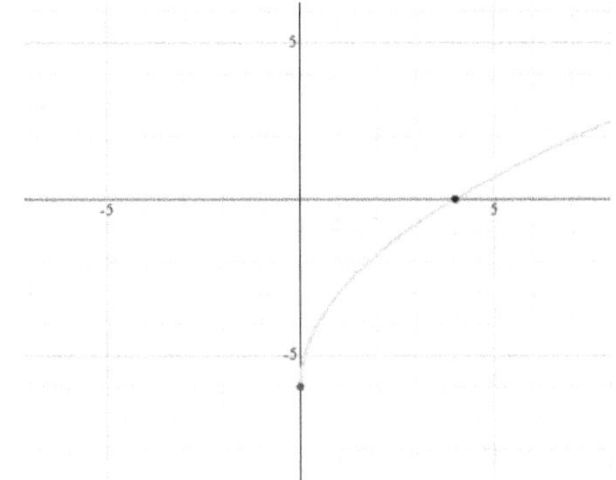

Solving radical equations

1) $\{81\}$
2) $\{36\}$
3) $\{16\}$
4) $\{\frac{1}{4}\}$
5) $\{321\}$
6) $\{43\}$
7) $\{19\}$
8) $\{54\}$
9) $\{148\}$
10) $\{59\}$
11) $\{5\}$
12) $\{32\}$
13) $\{6\}$
14) $\{72\}$
15) $\{194\}$
16) $\{0\}$
17) $\{20\}$
18) $\{5\}$
19) $\{2\}$
20) $\{9\}$
21) $\{2\}$
22) no solution
23) $\{-2\}$
24) $\{7\}$
25) $\{6\}$
26) $\{0\}$
27) $\{-4\}$
28) $\{-1\}$
29) $\{17\}$
30) $\{40\}$

Chapter 11 :

Sequences and Series

Topics that you'll practice in this chapter:

- ✓ Arithmetic Sequences
- ✓ Geometric Sequences
- ✓ Comparing Arithmetic and Geometric Sequences
- ✓ Finite Geometric Series
- ✓ Infinite Geometric Series

"The fact that people will be full of greed, fear, or folly is predictable. The sequence is not predictable." -Warren Buffett

Arithmetic Sequences

Find the next three terms of each arithmetic sequence.

1) $32, 26, 20, 14, 8, \ldots$

2) $-56, -44, -32, -20, \ldots$

3) $17, 26, 35, 44, 53, \ldots$

4) $5, 11, 17, 23, 29, \ldots$

Given the first term and the common difference of an arithmetic sequence find the first five terms and the explicit formula.

5) $a_1 = 20, d = 3$

6) $a_1 = -11, d = -5$

7) $a_1 = 32, d = 6$

8) $a_1 = 240, d = -80$

Given a term in an arithmetic sequence and the common difference find the first five terms and the explicit formula.

9) $a_{20} = -500, d = -50$

10) $a_{24} = 98, d = 7$

11) $a_{51} = -88.2, d = -5.2$

12) $a_{68} = -980, d = -27$

Given a term in an arithmetic sequence and the common difference find the recursive formula and the three terms in the sequence after the last one given.

13) $a_{21} = -187, d = -9$

14) $a_{12} = 63.5, d = 5.2$

15) $a_{31} = 58.2, d = 1.8$

16) $a_{42} = 6.8, d = 0.4$

Geometric Sequences

Determine if the sequence is geometric. If it is, find the common ratio.

1) 1, – 7, 49, – 343, …

2) – 3, – 12, – 48, – 192, …

3) 8, 24, 48, 240, …

4) – 5, – 10, – 20, – 40, …

Given the first term and the common ratio of a geometric sequence find the first five terms and the explicit formula.

5) $a_1 = 0.4, \mathrm{r} = -3$

6) $a_1 = 0.2, \mathrm{r} = 4$

Given the recursive formula for a geometric sequence find the common ratio, the first five terms, and the explicit formula.

7) $a_n = a_{n-1} \times 4, a_1 = 2$

8) $a_n = a_{n-1} . (-2), a_1 = -4$

9) $a_n = a_{n-1} . 5, a_1 = 0.2$

10) $a_n = a_{n-1} . 3, a_1 = -3$

Given two terms in a geometric sequence find the 6th term and the recursive formula.

11) $a_3 = 576$ and $a_5 = 36$

12) $a_2 = -0.4$ and $a_4 = -1.6$

Comparing Arithmetic and Geometric Sequences

For each sequence, state if it is arithmetic, geometric, or neither.

1) $6, 11, 16, 21, \ldots$

2) $2, 5, 8, 11, \ldots$

3) $3, 10, 20, 27, \ldots$

4) $1, 11, 22, 33, 44, \ldots$

5) $4, 8, 10, 24, 96, \ldots$

6) $2, 10, 20, 50, 200, \ldots$

7) $0.2, 1, 5, 25, 125, \ldots$

8) $4, 12, 36, 108, \ldots$

9) $-27, -34, -41, -48, -55, \ldots$

10) $-3, 15, -75, 375, -1875, \ldots$

11) $10, 25, 50, 65, 80, \ldots$

12) $5, 15, 150, 250, 350\ \ldots$

13) $-35, -20, -5, 10, 25, \ldots$

14) $a_n = 4 . 8^{n-1}$

15) $a_n = 3 . 6^{n-1}$

16) $a_n = 8 - 4n$

17) $a_n = -210 + 310n$

18) $a_n = 53 + 53n$

19) $a_n = -10 . (-5)^{n-1}$

20) $a_n = 49 + 63n$

21) $a_n = (3n)^4$

22) $a_n = 40 + 8n$

23) $a_n = -(13)^{n-1}$

24) $a_n = -10 . (0.2)^{n-1}$

25) $a_n = \frac{3n+2}{4^n}$

26) $a_n = \frac{47+13n}{7^n}$

27) $a_n = \frac{12-12n}{12^n}$

28) $a_n = \frac{32-a_{n-1}}{2n}$

29) $a_n = \frac{4}{15} - \frac{2}{7}n$

Finite Geometric Series

Evaluate the related series of each sequence.

1) $-2, 4, -8, 16$

2) $-1, 6, -36, 216, -1{,}296$

3) $-2, 8, -32, 128, -512$

4) $4, 8, 16, 32, 64$

5) $-4, -12, -36, -108$

6) $5, -15, 45, -135, 405$

Evaluate each geometric series described.

7) $1 + 5 + 25 + 125 \ldots, n = 5$ ____________

8) $1 - 6 + 36 - 216 \ldots, n = 6$ ____________

9) $-2 - 6 - 18 - 54 \ldots, n = 8$ ____________

10) $0.2 - 1 + 5 - 25 \ldots, n = 6$ ____________

11) $0.6 - 3.6 + 21.6 - 129.6 \ldots, n = 5$ ____________

12) $-1 - 4 - 16 - 64 \ldots, n = 6$ ____________

13) $a_1 = -2, r = 6, n = 5$ ____________

14) $a_1 = 1, r = 9, n = 6$ ____________

15) $\sum_{n=1}^{6} 4 \,.\, (-3)^{n-1}$ ________

16) $\sum_{n=1}^{7} 2 \,.\, (-5)^{n-1}$ ________

17) $\sum_{n=1}^{10} 0.1 \,.\, (2)^{n-1}$ ________

18) $\sum_{m=1}^{6} (-4)^{m-1}$ ________

19) $\sum_{m=1}^{6} 5 \times (3)^{m-1}$ ________

20) $\sum_{k=1}^{5} 7 \times (6)^{k-1}$ ________

Infinite Geometric Series

Determine if each geometric series converges or diverges.

1) $a_1 = -1.8,\ \ r = 6$

2) $a_1 = 10.8, r = 0.3$

3) $a_1 = -2, r = 6.1$

4) $a_1 = 5, r = 0.24$

5) $a_1 = 1.2, r = 8$

6) $-1, 6, -36, 216, \ldots$

7) $6, -1, \frac{1}{6}, -\frac{1}{36}, \frac{1}{216}, \ \ldots$

8) $512 + 64 + 8 + 1\ \ldots$

9) $-5 + \frac{15}{7} - \frac{45}{49} + \frac{135}{343}\ \ldots$

10) $\frac{400}{459} - \frac{200}{153} + \frac{100}{51} - \frac{50}{17}\ \ldots$

Evaluate each infinite geometric series described.

11) $a_1 = 2, r = -\frac{1}{4}$

12) $a_1 = 36, r = -\frac{1}{6}$

13) $a_1 = 9, r = \frac{1}{3}$

14) $a_1 = 12, r = \frac{1}{7}$

15) $1 + 0.2 + 0.04 + 0.008 + \cdots$

16) $64 - 16 + 4 - 1\ \ldots,$

17) $1 - 0.3 + 0.09 - 0.027\ \ldots,$

18) $-5 + \frac{15}{7} - \frac{45}{49} + \frac{135}{343}\ \ldots,$

19) $\sum_{k=1}^{\infty} 7^{k-1}$

20) $\sum_{i=1}^{\infty} (\frac{2}{5})^{i-1}$

21) $\sum_{k=1}^{\infty} (-\frac{2}{9})^{k-1}$

22) $\sum_{n=1}^{\infty} 6(\frac{1}{3})^{n-1}$

Answers of Worksheets

Arithmetic Sequences

1) $2, -4, -10$
2) $-8, 4, 16$
3) $62, 71, 80$
4) $35, 41, 47$
5) First Five Terms: $20, 23, 26, 29, 32$, Explicit: $a_n = 20 + 3(n-1)$
6) First Five Terms: $-11, -16, -21, -26, -31$, Explicit: $a_n = -11 - 5(n-1)$
7) First Five Terms: $32, 38, 44, 50, 56$, Explicit: $a_n = 32 + 6(n-1)$
8) First Five Terms: $240, 160, 80, 0, -80$, Explicit: $a_n = 240 - 80(n-1)$
9) First Five Terms: $450, 400, 350, 300, 250$, Explicit: $a_n = 450 - 50(n-1)$
10) First Five Terms: $-63, -56, -49, -42, -35$, Explicit: $a_n = -63 + 7(n-1)$
11) First Five Terms: $171.8, 166.6, 161.4, 156.2, 151$, Explicit: $a_n = 171.8 - 5.2(n-1)$
12) First Five Terms: $829, 802, 775, 748, 721$, Explicit: $a_n = 829 - 27(n-1)$
13) Next 3 terms: $-196, -205, -214$, Recursive: $a_n = a_{n-1} - 9, a_1 = -7$
14) Next 3 terms: $68.7, 73.9, 79.1, 84.3$ Recursive: $a_n = a_{n-1} + 5.2,\ a_1 = 6.3$
15) Next 3 terms: $60, 61.8, 63.6$, Recursive: $a_n = a_{n-1} + 1.8, a_1 = 4.2$
16) Next 3 terms: $7.2, 7.6, 8$, Recursive: $a_n = a_{n-1} + 0.4, a_1 = -9.6$

Geometric Sequences

1) $r = -7$
2) $r = 4$
3) not geometric
4) $r = 2$
5) First Five Terms: $0.4, -1.2, 3.6, -10.8, 32.4$

   Explicit: $a_n = 0.4 \times (-3)^{n-1}$
6) First Five Terms: $0.2, 0.8, 3.2, 12.8, 51.2$

   Explicit: $a_n = 0.2 \times (4)^{n-1}$
7) Common Ratio: $r = 4$

   First Five Terms: $2, 8, 32, 128, 512$

   Explicit: $a_n = 2 \,.\, (4)^{n-1}$
8) Common Ratio: $r = -2$

First Five Terms: $-4, 8, -16, 32, -64$

Explicit: $a_n = -4 \,.\, (-2)^{n-1}$

9) Common Ratio: $r = 5$

First Five Terms: $0.2, 1, 5, 25, 125, 625$

Explicit: $a_n = 0.2 \,.\, (5)^{n-1}$

10) Common Ratio: $r = 3$

First Five Terms: $-3, -9, -27, -81, -243$

Explicit: $a_n = -3 \,.\, (3)^{n-1}$

11) $a_6 = -9$, Recursive: $a_n = a_{n-1} \,.\, (\frac{-1}{4})$, $a_1 = 9{,}216$

12) $a_6 = -6.4$, Recursive: $a_n = a_{n-1} \,.\, (-2)$, $a_1 = 0.2$

Comparing Arithmetic and Geometric Sequences

1) Arithmetic
2) Arithmetic
3) Neither
4) Neither
5) Neither
6) Neither
7) Geometric
8) Geometric
9) Arithmetic
10) Geometric
11) Neither
12) Neither
13) Arithmetic
14) Geometric
15) Geometric
16) Arithmetic
17) Arithmetic
18) Arithmetic
19) Geometric
20) Arithmetic
21) Neither
22) Arithmetic
23) Geometric
24) Geometric
25) Neither
26) Neither
27) Neither
28) Neither
29) Arithmetic

Finite Geometric

1) 10
2) $-1{,}111$
3) -410
4) 124
5) -160
6) 305
7) 781
8) $-6{,}665$
9) $-6{,}560$
10) -520.8
11) 666.6
12) $-1{,}365$
13) $-3{,}110$
14) $66{,}430$
15) -728
16) $26{,}042$
17) 102.3
18) -819
19) $1{,}820$
20) $10{,}885$

Infinite Geometric

1) Diverges
2) Converges
3) Diverges
4) Converges
5) Diverges
6) Diverges
7) Converges
8) Converges
9) Converges
10) Converges
11) $\frac{8}{5}$
12) $\frac{216}{7}$
13) $\frac{27}{2}$
14) 14
15) $\frac{5}{4}$
16) $\frac{256}{5}$
17) $\frac{3}{4}$
18) $-\frac{7}{2}$
19) Infinite
20) $\frac{5}{3}$
21) $\frac{9}{11}$
22) 9

Chapter 12 :
Logarithms

Topics that you'll practice in this chapter:

- ✓ Rewriting Logarithms
- ✓ Evaluating Logarithms
- ✓ Properties of Logarithms
- ✓ Natural Logarithms
- ✓ Exponential Equations Requiring Logarithms
- ✓ Solving Logarithmic Equations

Mathematics is an art of human understanding.

— William Thurston

Rewriting Logarithms

Rewrite each equation in exponential form.

1) $\log_3 27 = 3$

2) $\log_2 128 = 7$

3) $\log_6 1{,}296 = 4$

4) $\log_5 625 = 4$

5) $\log_{11} 121 = 2$

6) $\log_{12} 1{,}728 = 3$

7) $\log_9 729 = 3$

8) $\log_3 729 = 6$

9) $\log_{10} 10{,}000 = 4$

10) $\log_7 343 = 3$

11) $\log_4 1{,}024 = 5$

12) $\log_{12} 144 = 2$

13) $\log_{13} 2{,}197 = 3$

14) $\log_{25} 5 = \frac{1}{2}$

15) $\log_{81} 3 = \frac{1}{4}$

16) $\log_{3,125} 5 = \frac{1}{5}$

17) $\log_{1,000} 10 = \frac{1}{3}$

18) $\log_5 \frac{1}{125} = -3$

19) $\log_4 \frac{1}{16} = -2$

20) $\log_a \frac{7}{4} = b$

Rewrite each exponential equation in logarithmic form.

21) $2^5 = 32$

22) $4^3 = 64$

23) $5^4 = 625$

24) $11^3 = 1{,}331$

25) $3^5 = 243$

26) $6^4 = 1{,}296$

27) $7^4 = 2{,}401$

28) $9^3 = 729$

29) $4^{-5} = \frac{1}{1{,}024}$

30) $3^{-8} = \frac{1}{6{,}561}$

31) $11^{-2} = \frac{1}{121}$

32) $12^{-3} = \frac{1}{1{,}728}$

33) $4^{-5} = \frac{1}{1{,}024}$

34) $10^{-5} = \frac{1}{100{,}000}$

Evaluating Logarithms

Evaluate each logarithm.

1) $\log_3 729 =$

2) $\log_2 256 =$

3) $\log_3 243 =$

4) $\log_4 64 =$

5) $\log_8 64 =$

6) $\log_{11} 121 =$

7) $\log_{10} 10{,}000 =$

8) $\log_5 \frac{1}{25} =$

9) $\log_4 \frac{1}{256} =$

10) $\log_2 \frac{1}{32} =$

11) $\log_6 \frac{1}{36} =$

12) $\log_9 \frac{1}{81} =$

13) $\log_{12} \frac{1}{144} =$

14) $\log_{1,000} \frac{1}{10} =$

15) $\log_{243} 3 =$

16) $\log_4 \frac{1}{16} =$

17) $\log_8 \frac{1}{512} =$

18) $\log_3 \frac{1}{81} =$

Circle the points which are on the graph of the given logarithmic functions.

19) $y = 4log_4(3x - 2) + 1$ $(3, 4)$, $(2, 5)$, $(7, 4)$

20) $y = 5log_6(12x) - 7$ $(2, -2)$, $(\frac{1}{3}, 12)$, $(\frac{1}{2}, -2)$

21) $y = -2log_3 9(x - 5) + 5$ $(6, -3)$, $(8, -1)$, $(1, 6)$

22) $y = \frac{1}{4}log_6(6x) + \frac{1}{2}$ $(6, 1)$, $(6, \frac{1}{4})$, $(4, \frac{1}{4})$

23) $y = -2log_8 8(x + 4) + 9$ $(-4, 0)$, $(0, 9)$, $(-2, 6\frac{1}{3})$

24) $y = -log_5(x + 15) - 6$ $(10, -\frac{1}{5})$, $(10, -8)$, $(11, -\frac{2}{5})$

25) $y = -3log_2(2x + 6) + 7$ $(5, -5)$, $(-5, -5)$, $(-2, -2)$

Properties of Logarithms

Expand each logarithm.

1) $\log(11 \times 4) =$

2) $\log(13 \times 5) =$

3) $\log(4 \times 12) =$

4) $\log(\frac{2}{7}) =$

5) $\log(\frac{4}{9}) =$

6) $\log(\frac{5}{8})^3 =$

7) $\log(6 \times 5^4) =$

8) $\log(\frac{14}{3})^5 =$

9) $\log\left(\frac{3^4}{8}\right) =$

10) $\log(x \times y)^8 =$

11) $\log(x^6 \times y^{12} \times z^2) =$

12) $\log\left(\frac{u^8}{v^3}\right) =$

13) $\log\left(\frac{x}{y^7}\right) =$

Condense each expression to a single logarithm.

14) $\log 8 - \log 13 =$

15) $\log 6 + \log 11 =$

16) $4\log 2 - 7\log 5 =$

17) $10\log 4 - 3\log 7 =$

18) $3\log 9 - \log 17 =$

19) $11\log 6 - 9\log 4 =$

20) $\log 15 - 6\log 7 =$

21) $6\log 8 + 4\log 10 =$

22) $12\log 5 + 14\log 9 =$

23) $17\log_8 a + 6\log_8 b =$

24) $2\log_9 x - 3\log_9 y =$

25) $\log_{11} u - 16\log_{11} v =$

26) $8\log_{15} u + 9\log_{15} v =$

27) $32\log_6 u - 25\log_6 v =$

Natural Logarithms

Solve each equation for x.

1) $e^x = 9$
2) $e^x = 36$
3) $e^x = 49$
4) $\ln x = 3$
5) $\ln(\ln x) = 7$
6) $e^x = 4$
7) $\ln(5x + 2) = 1$
8) $\ln(7x + 4) = 3$
9) $\ln(9x + 5) = 4$
10) $\ln x = \frac{1}{9}$
11) $\ln 11x = e^5$
12) $\ln x = \ln 6 + \ln 7$
13) $\ln x = 4\ln 3 + \ln 2$

Evaluate without using a calculator.

14) $11\ln e =$
15) $\ln e^{10} =$
16) $4 \ln e =$
17) $\ln e^{21} =$
18) $32\ln e =$
19) $4\ln e^5 =$
20) $e^{\ln 22} =$
21) $e^{3\ln 3} =$
22) $e^{3\ln 5} =$
23) $\ln \sqrt[11]{e} =$

Reduce the following expressions to simplest form.

24) $e^{-4\ln 9 + 4\ln 3} =$
25) $e^{-3\ln\left(\frac{5}{4e}\right)} =$
26) $2\ln(e^4) =$
27) $\ln\left(\frac{1}{e}\right)^4 =$
28) $e^{\ln 9 + 3\ln 3} =$
29) $e^{\ln\left(\frac{13}{e}\right)} =$
30) $8\ln(1^{-3e}) =$
31) $2\ln\left(\frac{1}{e}\right)^{-3} =$
32) $6\ln\left(\frac{\sqrt[3]{e}}{3e}\right) =$
33) $e^{-4\ln e + 2\ln 5} =$
34) $e^{\ln\frac{4}{e}} =$
35) $11\ln(e^e) =$

Exponential Equations and Logarithms

Solve each equation for the unknown variable.

1) $3^{4n} = 243$
2) $5^{3r} = 625$
3) $6^{2n-1} = 216$
4) $16^{2r+3} = 4$
5) $169^{2x} = 13$
6) $7^{-3v-3} = 49$
7) $2^{4n} = 128$
8) $11^{n-1} = 1{,}331$
9) $\frac{9^{3a}}{3^{2a}} = 729$
10) $13^5 \times 13^{-4v} = 169$
11) $4^{3n} = \frac{1}{64}$
12) $(\frac{1}{11})^{2n} = 121$
13) $2{,}187^{3x} = 3$
14) $13^{5-7x} = 13^{-2x}$
15) $11^{-3x} = 11^{2x-7}$
16) $3^{5n} = 243$
17) $17^{5x+3} = 17^{6x}$
18) $15^{3n} = 225$
19) $4^{-3k} = 512$
20) $8^{-4r} = 8^{-5r+2}$
21) $8^{2x+3} = 8^{5x}$
22) $10^{3x-2} = 100{,}000$
23) $16 \times 64^{-v} = 128$
24) $\frac{128}{2^{-3m}} = 2^{4m+5}$
25) $14^{-5n} \times 14^{2n+3} = 14^{-2n}$
26) $(\frac{1}{9})^{4n+3} \times (\frac{1}{9})^{-3n-8} = (\frac{1}{9})^{-4n}$

Solve each problem. (Round to the nearest whole number)

27) A substance decays 16% each day. After 8 days, there are 6 milligrams of the substance remaining. How many milligrams were there initially? ____________

28) A culture of bacteria grows continuously. The culture doubles every 4 hours. If the initial number of bacteria is 20, how many bacteria will there be in 13 hours?

____

29) Bob plans to invest \$11,200 at an annual rate of 3.5%. How much will Bob have in the account after three years if the balance is compounded quarterly? ______

30) Suppose you plan to invest \$8,000 at an annual rate of 5%. How much will you have in the account after 6 years if the balance is compounded monthly? ______

Solving Logarithmic Equations

Find the value of the variables in each equation.

1) $2\log(x) + 5 = 9$

2) $\log_4 4x + 3 = 5$

3) $-\log_8(8x) + 2 = 3$

4) $\log 2x - \log 4 = 1$

5) $\log 5x + \log 25 = 1$

6) $\log 4 - \log x = 3$

7) $\log 4x + \log 2 = \log 16$

8) $-6\log_3(5x - 1) = -36$

9) $\log 4x = \log(8x - 1)$

10) $\log(4k - 6) = \log(k - 3)$

11) $\log(5p + 2) = \log(p + 4)$

12) $-30 + \log_4(3n + 2) = -30$

13) $\log_4(4x - 4) = \log_4(x^2)$

14) $\log_8(k^2 + 15) = \log_8(-6k - 3)$

15) $\log(16 + 6b) = \log(10b^2 + 12b)$

16) $\log_6(2x + 5) - \log_6 x = \log_6 9$

17) $\log_5 5 + \log_5(x^2 + 1) = \log_5 25$

18) $\log_6(x + 3) + \log_6(x + 1) = \log_6 8$

Find the value of x in each natural logarithm equation.

19) $\ln 8 - \ln(4x + 8) = 4$

20) $\ln(x + 5) - \ln(x + 2) = \ln 10$

21) $\ln e^6 - \ln(x - 1) = 3$

22) $\ln(2x - 8) + \ln(x - 4) = \ln 8$

23) $\ln 5x - \ln(x + 4) = \ln 2$

24) $\ln(8x - 4) - \ln(x - 2) = \ln 25$

25) $\ln(3x + 2) - 4\ln 2 = 5$

26) $\ln(2x - 5) + \ln(x - 3) = \ln 6$

27) $\ln(x - 1) + \ln(4x - 7) = \ln(7)$

28) $2\ln 3x - \ln(x + 10) = \ln 2x$

29) $\ln x^4 + \ln x^8 = 4\ln(2x)$

30) $\ln x^{10} - \ln(x^2 + 10) = 10\ln 2x$

31) $8\ln(x - 2) = 4\ln(x^2 - 4x + 4)$

32) $\ln(x^4 + 10) = \ln(x^2 + 9)$

33) $2\ln x - 2\ln(x + 8) = \ln(x^2)$

34) $\ln(2x + 1) - \ln(4x + 1) = \ln 4$

35) $\ln 16 + 2\ln(x - 2) = \ln 4$

36) $\ln e^2 + \ln(5x - 6) = \ln(5) + 3$

Answers of Worksheets

Rewriting Logarithms

1) $3^3 = 27$
2) $2^7 = 128$
3) $6^4 = 1{,}296$
4) $5^4 = 625$
5) $11^2 = 121$
6) $12^3 = 1{,}728$
7) $9^3 = 729$
8) $3^6 = 729$
9) $10^4 = 10{,}000$
10) $7^3 = 343$
11) $4^5 = 1{,}024$
12) $12^2 = 144$
13) $13^3 = 2{,}197$
14) $25^{\frac{1}{2}} = 5$
15) $81^{\frac{1}{4}} = 3$
16) $3{,}125^{\frac{1}{5}} = 5$
17) $1{,}000^{\frac{1}{3}} = 10$
18) $5^{-3} = \frac{1}{125}$
19) $4^{-2} = \frac{1}{16}$
20) $a^b = \frac{7}{4}$
21) $\log_2 32 = 5$
22) $\log_4 64 = 3$
23) $\log_5 625 = 4$
24) $\log_{11} 1{,}331 = 3$
25) $\log_3 243 = 5$
26) $\log_6 1{,}296 = 4$
27) $\log_7 2{,}401 = 4$
28) $\log_9 729 = 3$
29) $\log_4 \frac{1}{1{,}024} = -5$
30) $\log_3 \frac{1}{6{,}561} = -8$
31) $\log_{11} \frac{1}{121} = -2$
32) $\log_{12} \frac{1}{1{,}728} = -3$
33) $\log_4 \frac{1}{1{,}024} = -5$
34) $\log_{10} \frac{1}{100{,}000} = -5$

Evaluating Logarithms

1) 6
2) 8
3) 5
4) 3
5) 2
6) 2
7) 4
8) -2
9) -4
10) -5
11) -2
12) -2
13) -2
14) $-\frac{1}{3}$
15) $\frac{1}{5}$
16) -2
17) -3
18) -4
19) $(2, 5)$
20) $(\frac{1}{2}, -2)$
21) $(8, -1)$
22) $(6, 1)$
23) $(-2, 6\frac{1}{3})$
24) $(10, -8)$
25) $(5, -5)$

Properties of Logarithms

1) $\log 11 + \log 4$
2) $\log 13 + \log 5$
3) $\log 4 + \log 12$
4) $\log 2 - \log 7$
5) $\log 4 - \log 9$
6) $3 \log 5 - 3 \log 8$
7) $\log 6 + 4 \log 5$
8) $5\log 14 - 5 \log 3$

9) $4 \log 3 - \log 8$
10) $8 \log x + 8 \log y$
11) $6\log x + 12\log y + 2\log z$
12) $8\log u - 3\log v$
13) $\log x - 7\log y$
14) $\log \frac{8}{13}$
15) $\log(6 \times 11)$
16) $\log \frac{2^4}{5^7}$
17) $\log \frac{4^{10}}{7^3}$
18) $\log \frac{9^3}{17}$
19) $\log \frac{6^{11}}{4^9}$
20) $\log \frac{15}{7^6}$
21) $\log (8^6 \times 10^4)$
22) $\log (5^{12} \times 9^{14})$
23) $\log_8 (a^{17}b^6)$
24) $\log_9 \frac{x^2}{y^3}$
25) $\log_{11} \frac{u}{v^{16}}$
26) $\log_{15}(u^8 \times v^9)$
27) $\log_6 \frac{u^{32}}{v^{25}}$

Natural Logarithms

1) $x = ln\, 9$
2) $x = \ln 36, x = 2\ln\ (6)$
3) $x = \ln 49, x = 2\ln\ (7)$
4) $x = e^3$
5) $x = e^{e^7}$
6) $x = ln\, 4$
7) $x = \frac{e-2}{5}$
8) $x = \frac{e^3-4}{7}$
9) $x = \frac{e^4-5}{9}$
10) $x = \sqrt[9]{e}$
11) $x = \frac{e\, e^5}{11}$
12) $x = 42$
13) $x = 162$
14) 11
15) 10
16) 4
17) 21
18) 32
19) 20
20) 22
21) 27
22) 125
23) $\frac{1}{11}$
24) $\frac{1}{81}$
25) $\frac{64e^3}{125}$
26) 8
27) -4
28) 243
29) $\frac{13}{e}$
30) 0
31) 6
32) $\ln\left(\frac{1}{3^6e^4}\right) = -10.6$
33) $25e^{-4} = \frac{25}{e^4}$
34) $\frac{4}{e}$
35) $11e$

Exponential Equations and Logarithms

1) $\frac{5}{4}$
2) $\frac{4}{3}$
3) 2
4) $-\frac{5}{4}$
5) $\frac{1}{4}$
6) $-\frac{5}{3}$
7) $\frac{7}{4}$
8) 4

9) $\frac{3}{2}$

10) $\frac{3}{4}$

11) -1

12) -1

13) $\frac{1}{21}$

14) 1

15) $\frac{7}{5}$

16) 1

17) 3

18) $\frac{2}{3}$

19) $-\frac{3}{2}$

20) 2

21) 1

22) $\frac{7}{3}$

23) $-\frac{1}{2}$

24) 2

25) 3

26) 1

27) 24.2

28) 190.27

29) \$12,432.4

30) \$10,792.14

Solving Logarithmic Equations

1) $\{100\}$

2) $\{4\}$

3) $\{\frac{1}{64}\}$

4) $\{20\}$

5) $\{\frac{2}{25}\}$

6) $\{\frac{1}{250}\}$

7) $\{2\}$

8) $\{146\}$

9) $\{\frac{1}{4}\}$

10) No Solution

11) $\{\frac{1}{2}\}$

12) $\{-\frac{1}{3}\}$

13) $\{2\}$

14) No Solution

15) $\{1, -\frac{8}{5}\}$

16) $\{\frac{5}{7}\}$

17) $\{2, -2\}$

18) $\{1\}$

19) $x = \frac{8-8e^4}{4e^4}$

20) $\{-\frac{5}{3}\}$

21) $e^3 + 1$

22) $\{6\}$

23) $\{\frac{8}{3}\}$

24) $\{\frac{46}{17}\}$

25) $x = \frac{16e^5-2}{3}$

26) $x = \frac{9}{2}$

27) $x = \frac{11}{4}$

28) $x = \frac{20}{7}$

29) $e^{\frac{\ln(2)}{2}}$

30) No Solution

31) $x > 2$

32) No Solution

33) No Solution

34) $x = -\frac{3}{14}$

35) $x = \frac{5}{2}$

36) $x = \frac{5e+6}{5}$

Chapter 13 :

Geometry and Solid Figures

Topics that you'll practice in this chapter:

- ✓ Angles
- ✓ Pythagorean Relationship
- ✓ Triangles
- ✓ Polygons
- ✓ Trapezoids
- ✓ Circles
- ✓ Cubes
- ✓ Rectangular Prism
- ✓ Cylinder
- ✓ Pyramids and Cone

Geometry is the archetype of the beauty of the world.

Johannes Kepler

Angles

What is the value of x in the following figures?

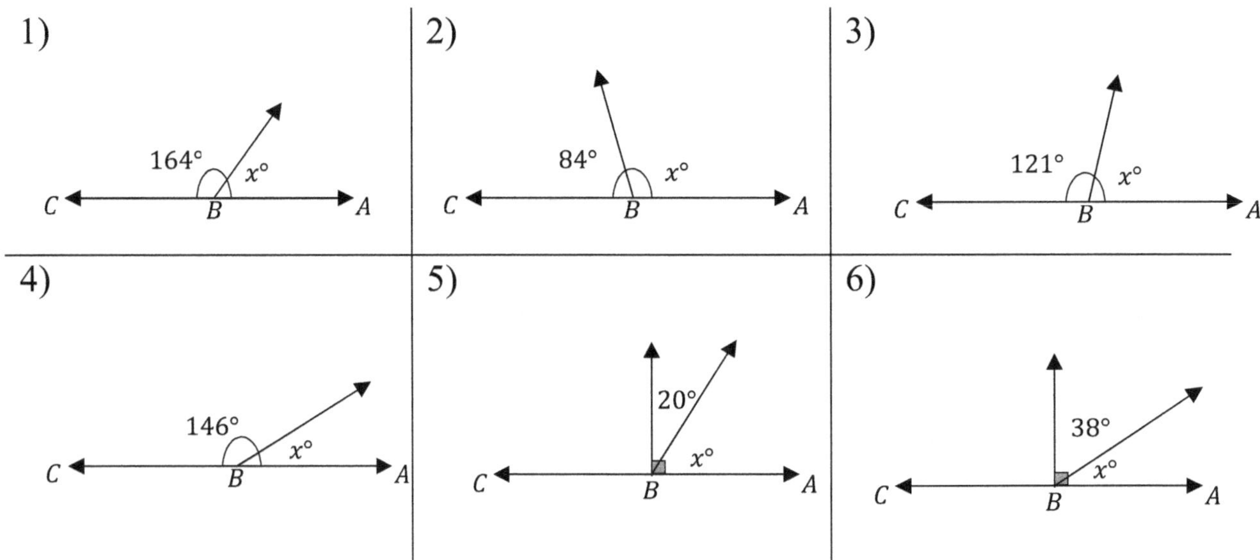

Calculate.

7) Two supplement angles have equal measures. What is the measure of each angle? ____________________

8) The measure of an angle is seven fifth the measure of its supplement. What is the measure of the angle? ____________________

9) Two angles are complementary and the measure of one angle is 24 less than the other. What is the measure of the smaller angle? ____________________

10) Two angles are complementary. The measure of one angle is one fifth the measure of the other. What is the measure of the bigger angle? ____________________

11) Two supplementary angles are given. The measure of one angle is 40° less than the measure of the other. What does the smaller angle measure? ____________________

Pythagorean Relationship

 **Do the following lengths form a right triangle?**

1)

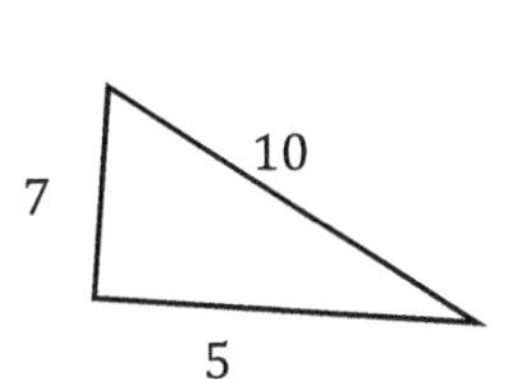

2)

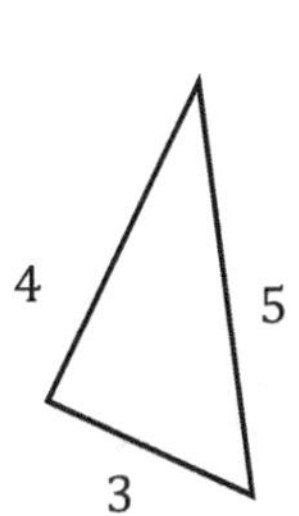

3)

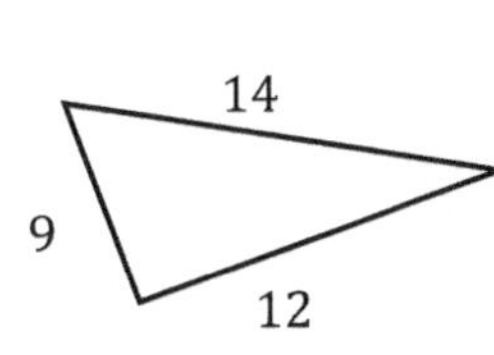

4)

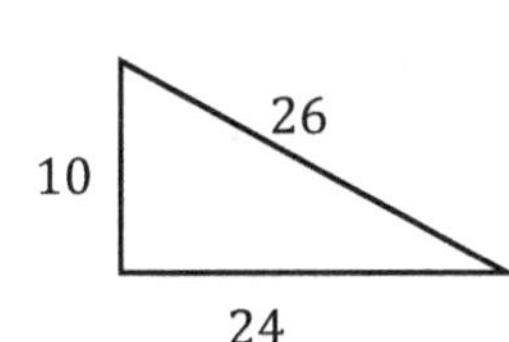

5)

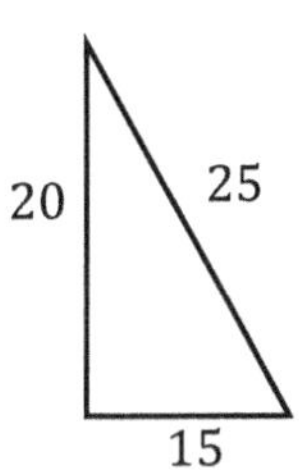

6)

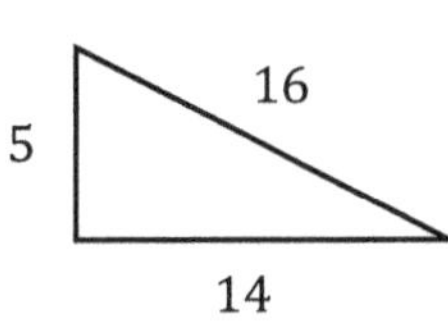

7)

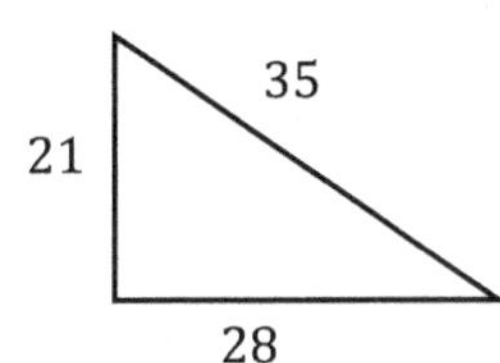

8)

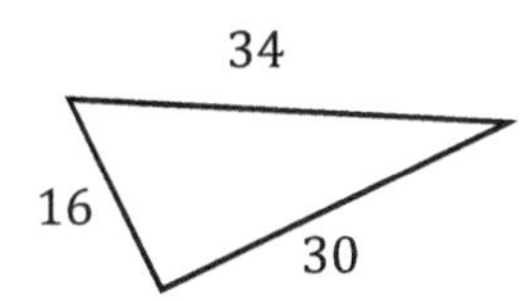

Find the missing side?

9)

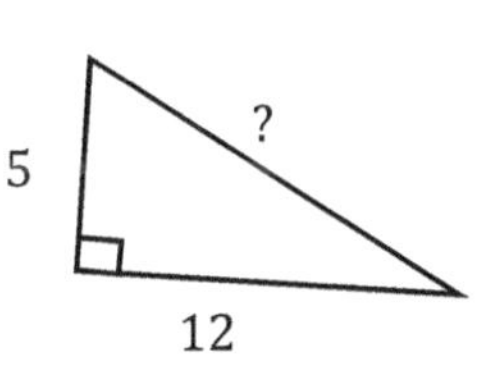

10)

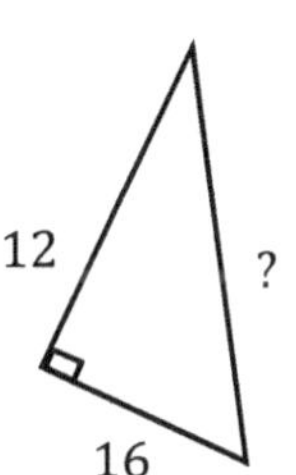

11)

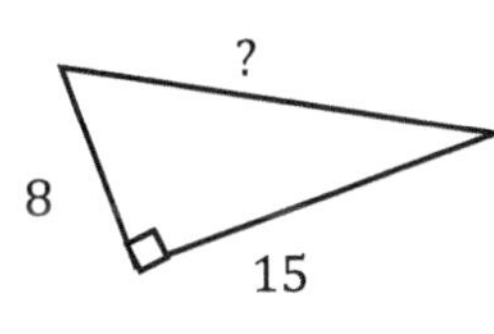

12)

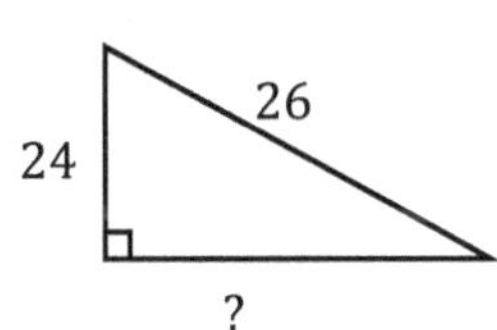

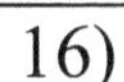

13)

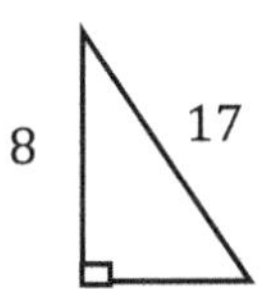

14)

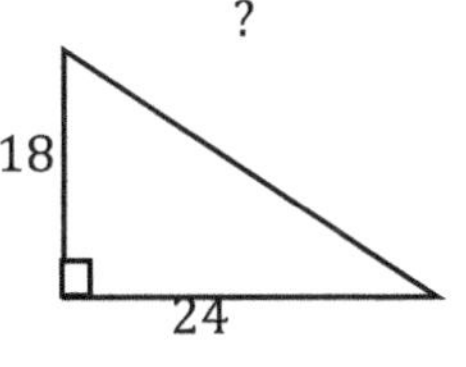

15)

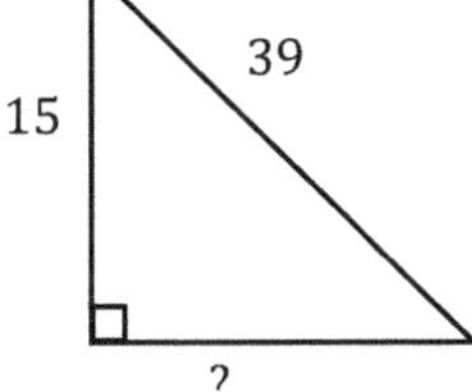

16)

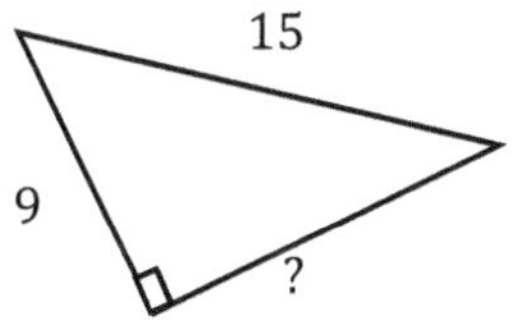

Triangles

Find the measure of the unknown angle in each triangle.

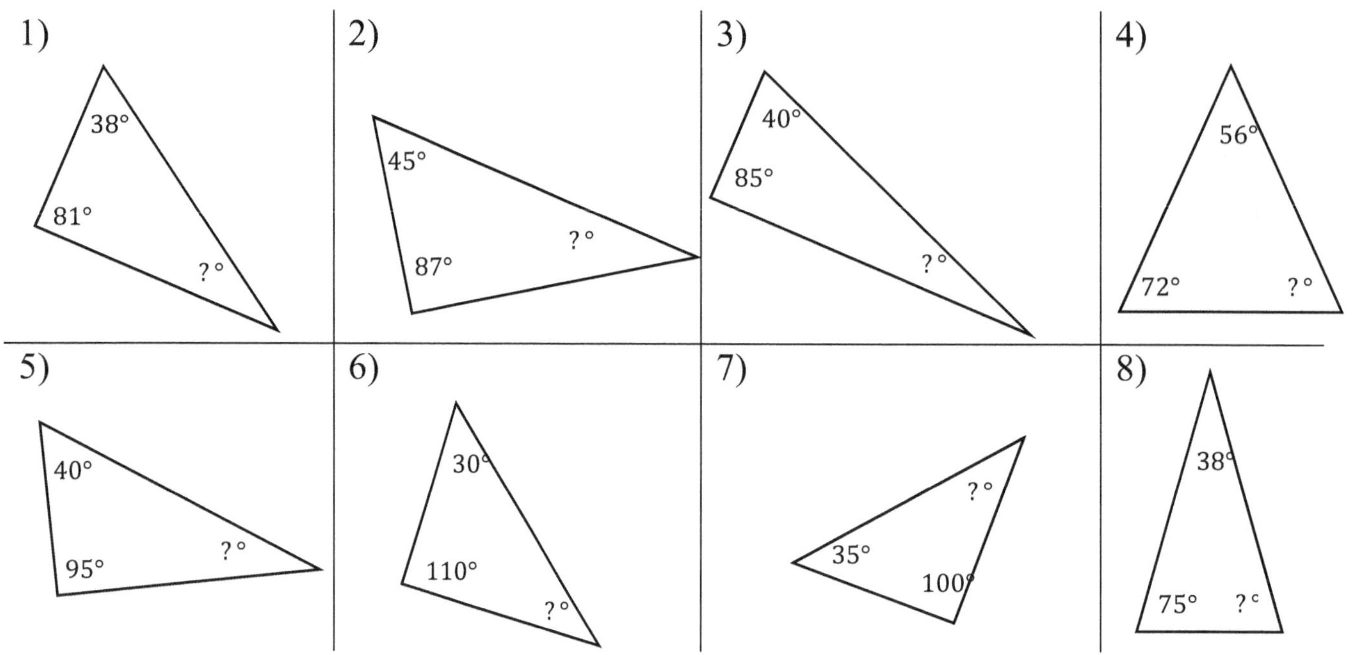

Find area of each triangle.

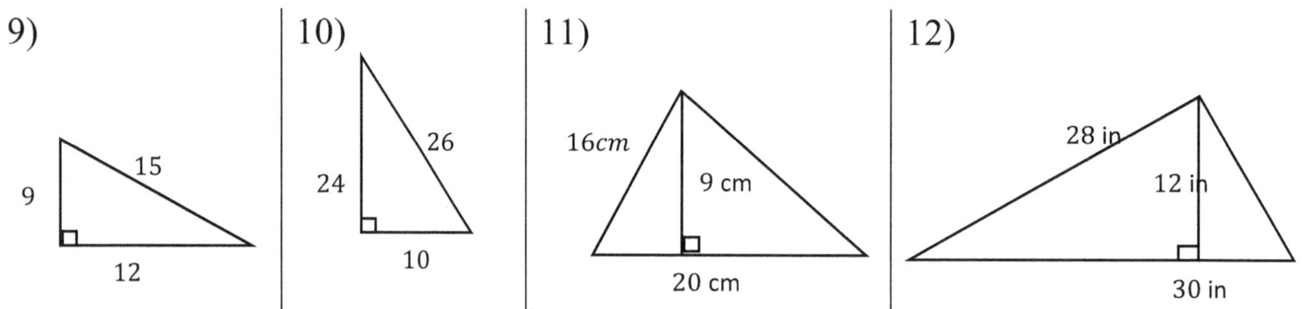

Polygons

Find the perimeter of each shape.

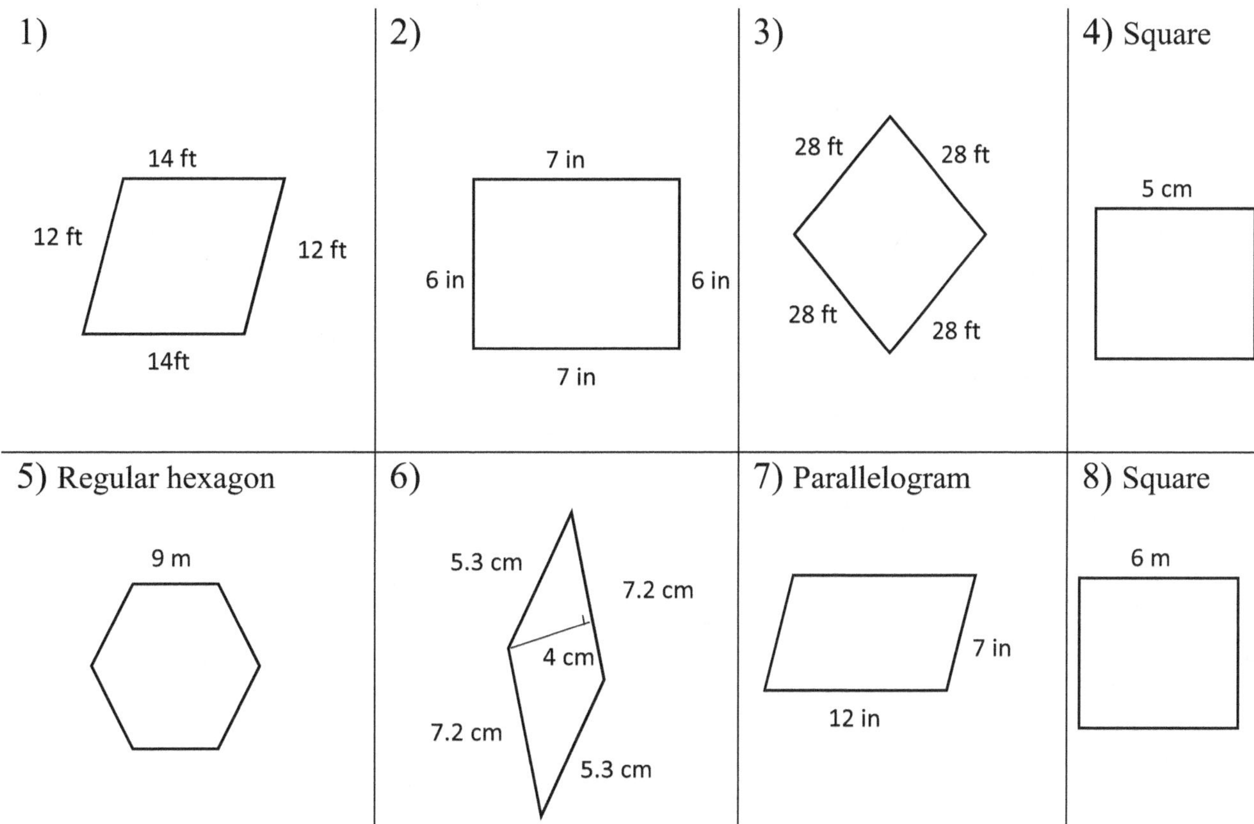

Find the area of each shape.

9) Parallelogram

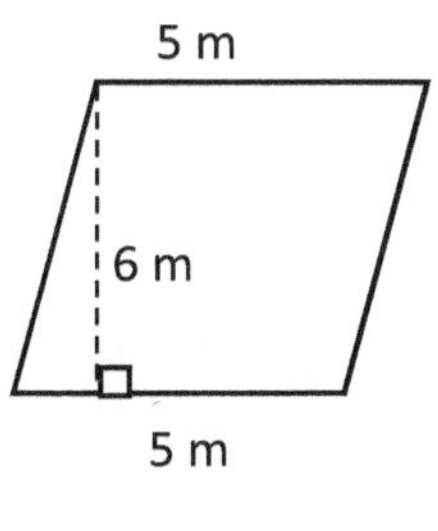

10) Rectangle

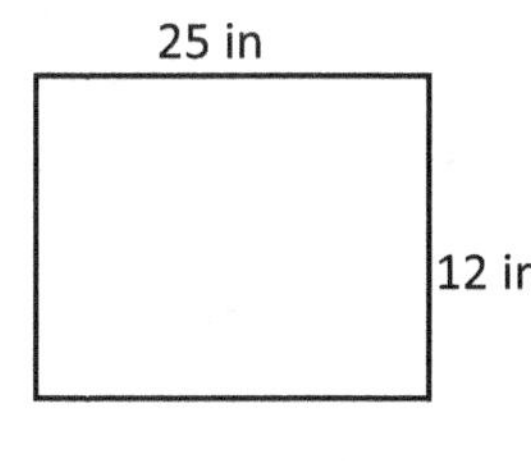

11) Rectangle

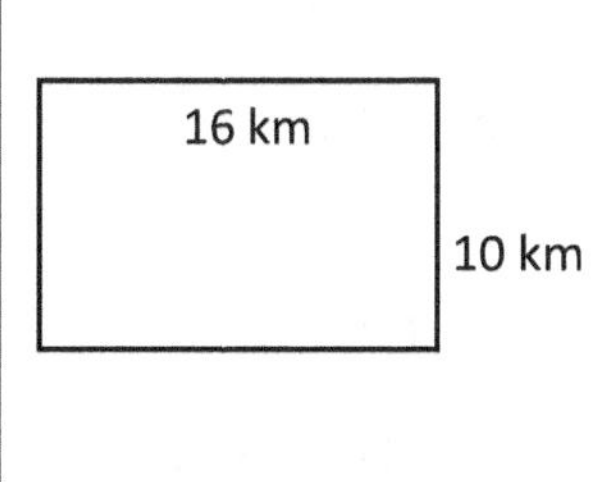

12) Square

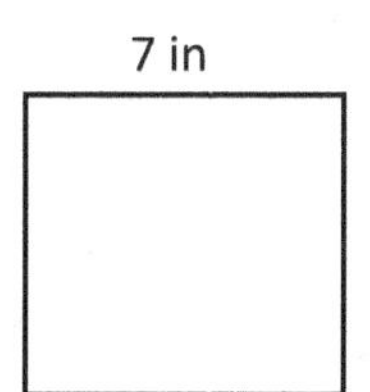

Trapezoids

Find the area of each trapezoid.

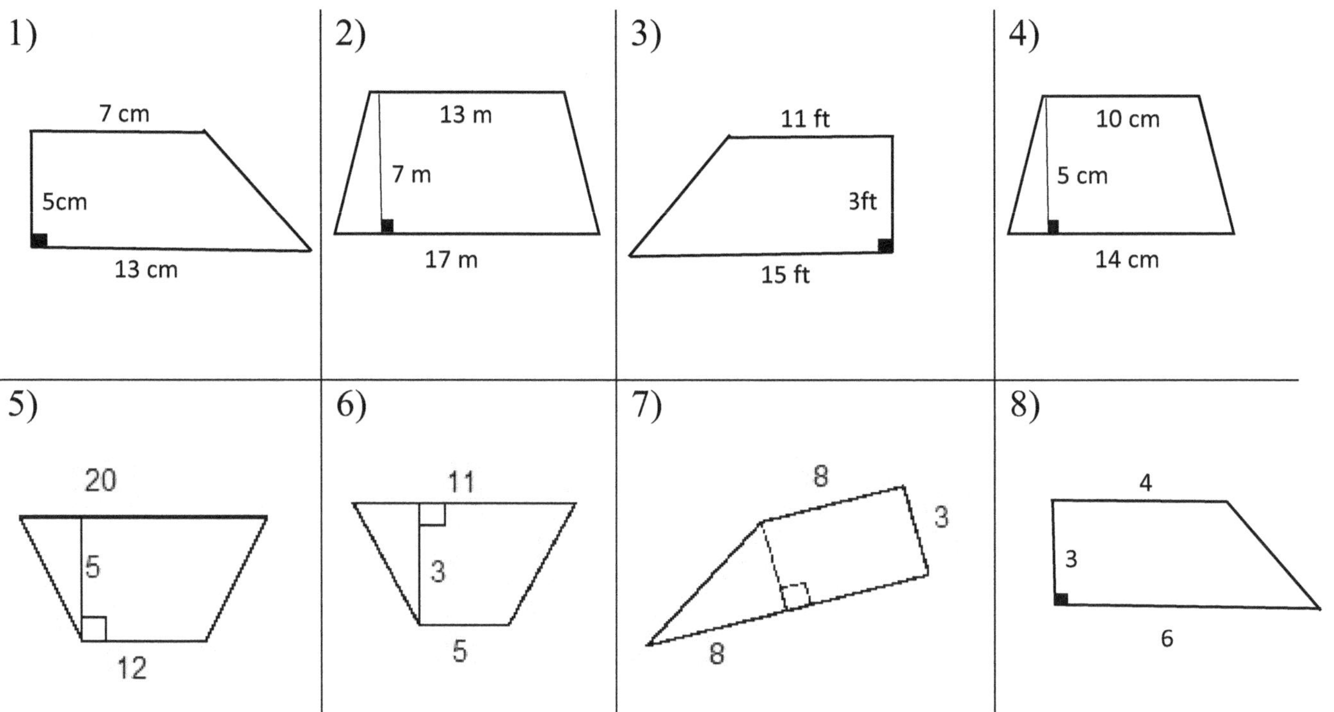

Calculate.

1) A trapezoid has an area of 45 cm^2 and its height is 5 cm and one base is 5 cm. What is the other base length? ______________________

2) If a trapezoid has an area of 99 ft^2 and the lengths of the bases are 8 ft and 10 ft, find the height? ______________________

3) If a trapezoid has an area of 126 m^2 and its height is 14 m and one base is 6 m, find the other base length? ______________________

4) The area of a trapezoid is 440 ft^2 and its height is 22 ft. If one base of the trapezoid is 15 ft, what is the other base length?

____

Circles

Find the area of each circle. ($\pi = 3.14$)

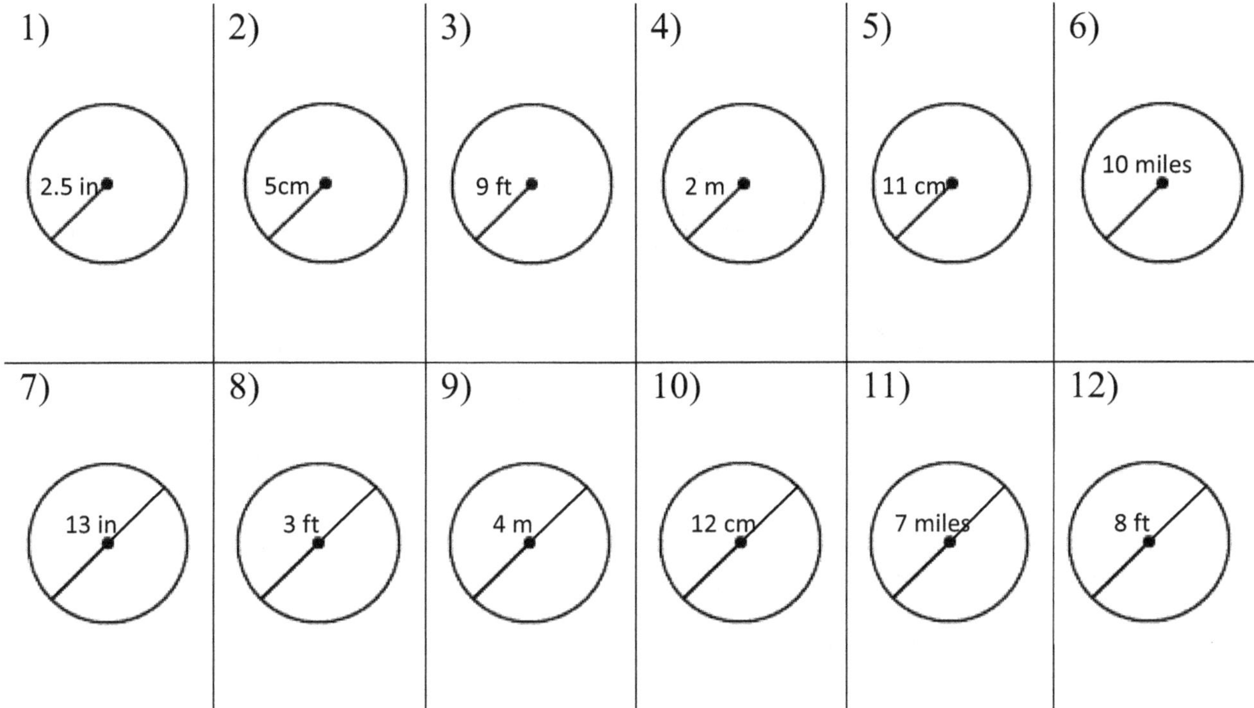

Complete the table below. ($\pi = 3.14$)

Circle No.	Radius	Diameter	Circumference	Area
1	$1\ in$	$2\ in$	$6.28\ in$	$3.14\ in^2$
2		$10\ m$		
3				$28.26\ ft^2$
4			$47.1\ mi$	
5		$11\ km$		
6	$7\ cm$			
7		$12\ ft$		
8				$314\ m^2$
9			$56.52\ in$	
10	$4.5\ ft$			

Cubes

Find the volume of each cube.

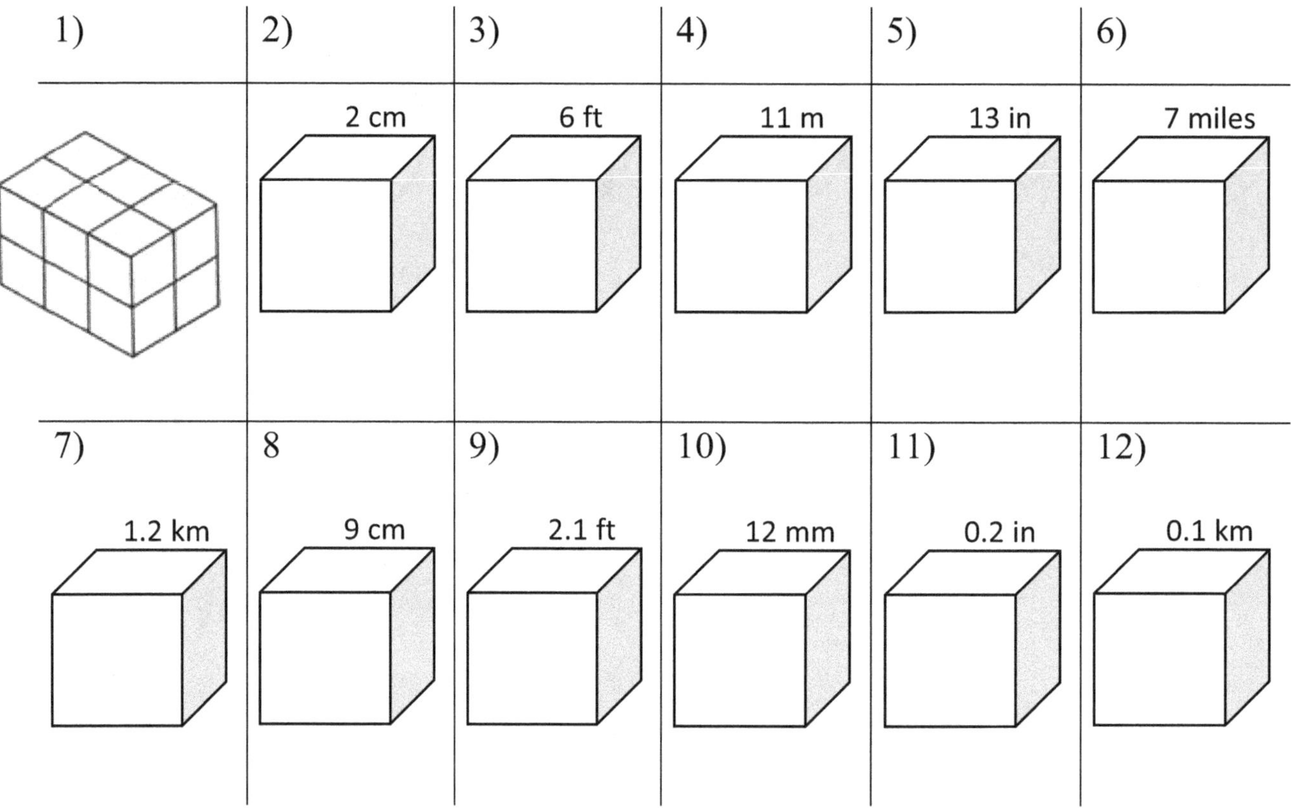

Find the surface area of each cube.

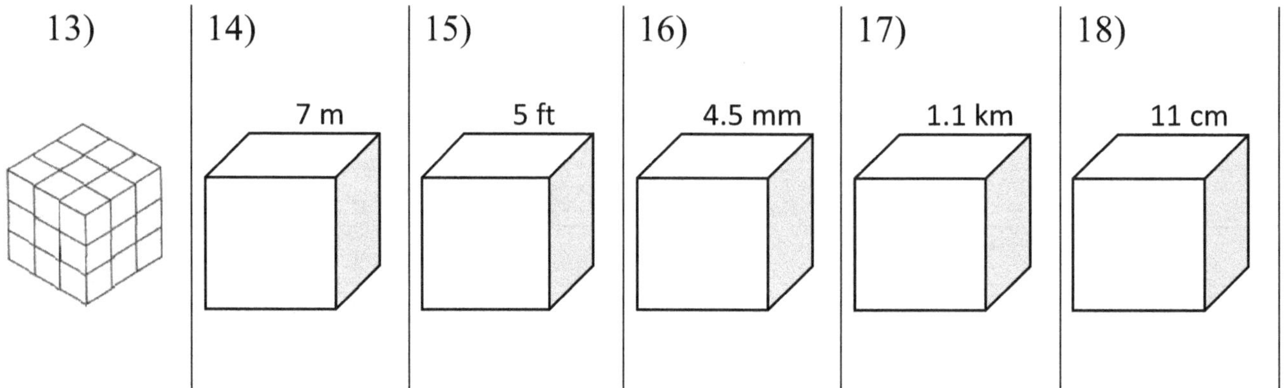

Rectangular Prism

Find the volume of each Rectangular Prism.

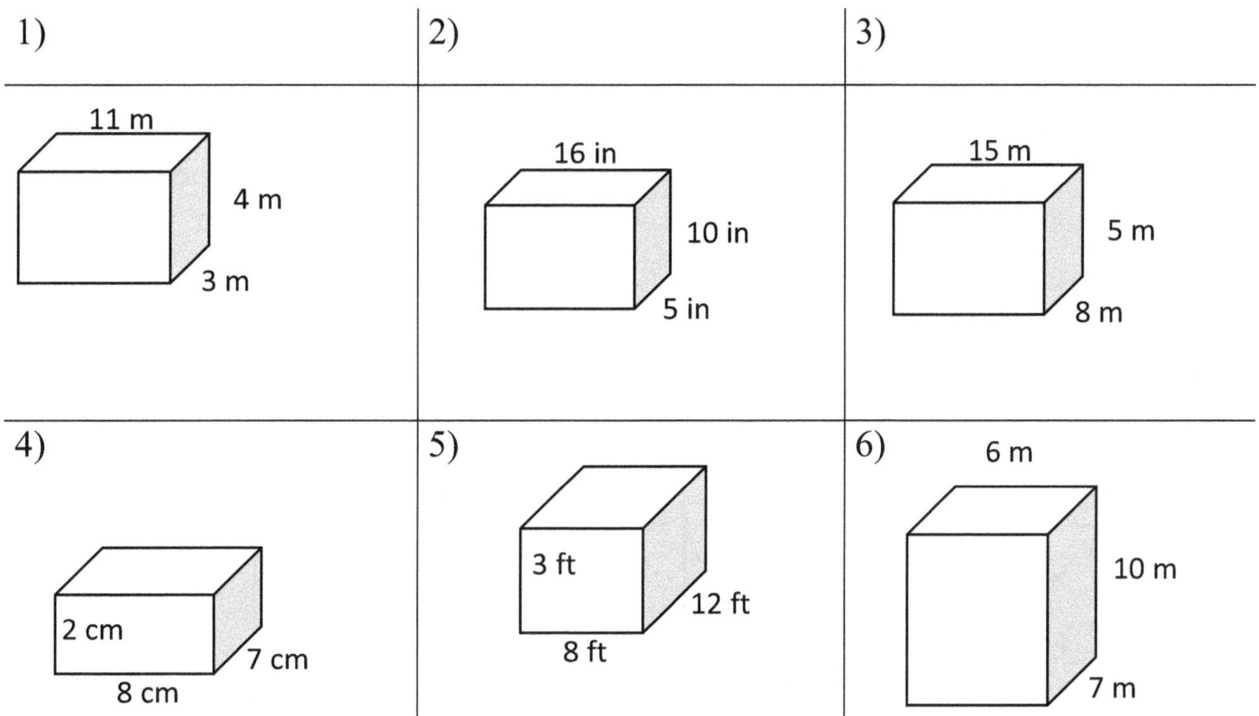

Find the surface area of each Rectangular Prism.

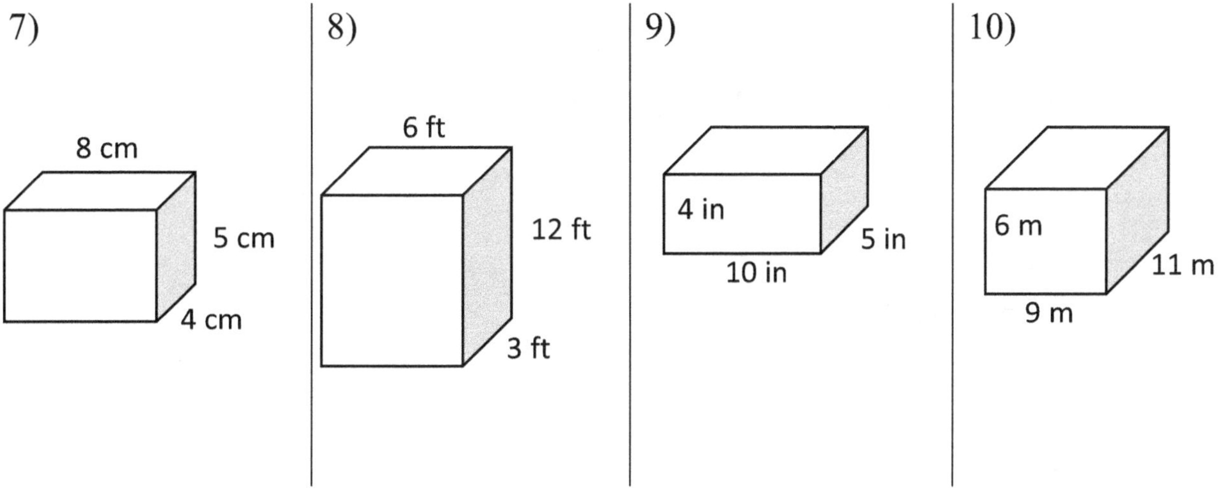

Cylinder

Find the volume of each Cylinder. Round your answer to the nearest tenth. ($\pi = 3.14$)

1)

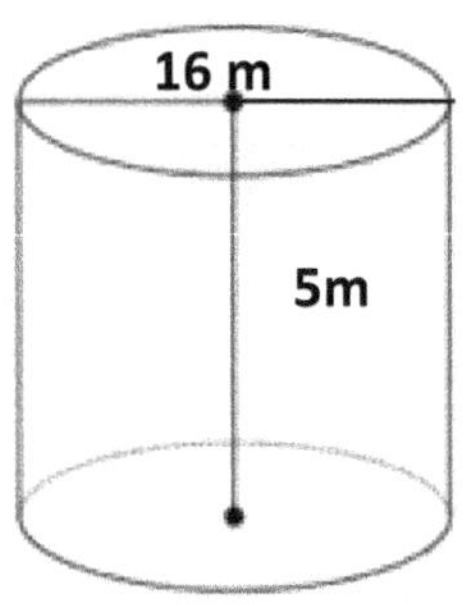

2)

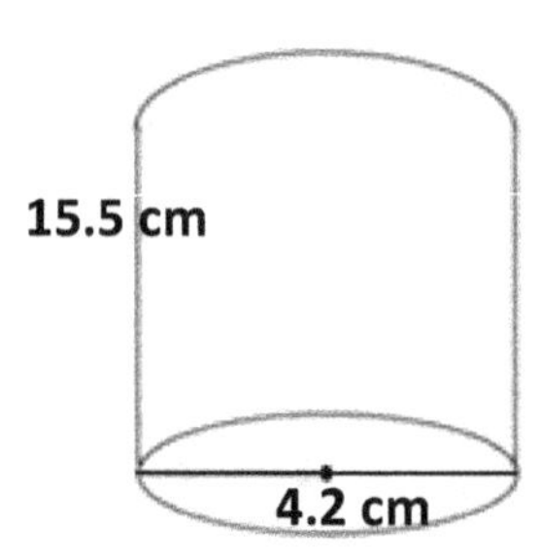

3)

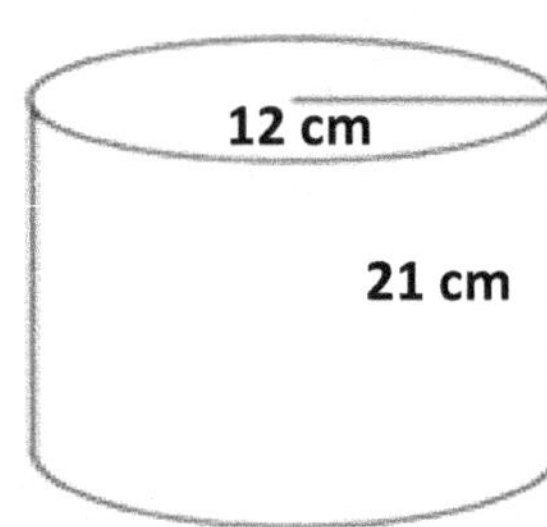

4)

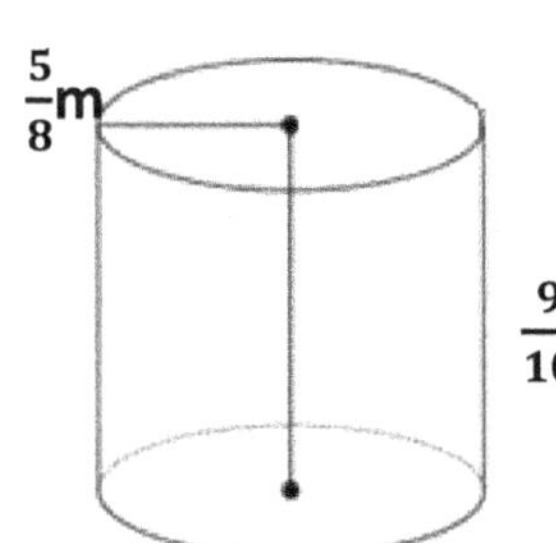

5)

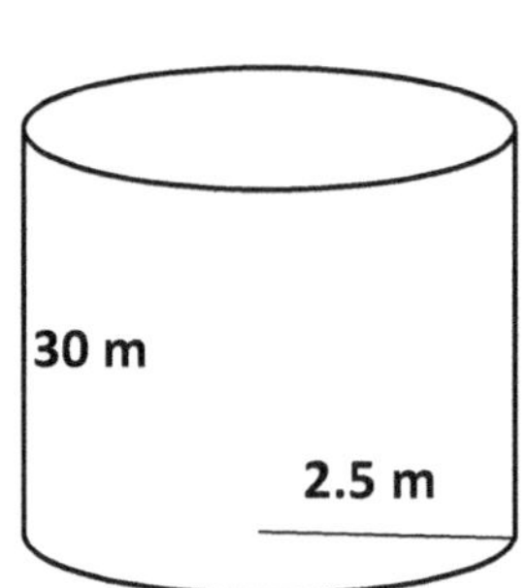

6)

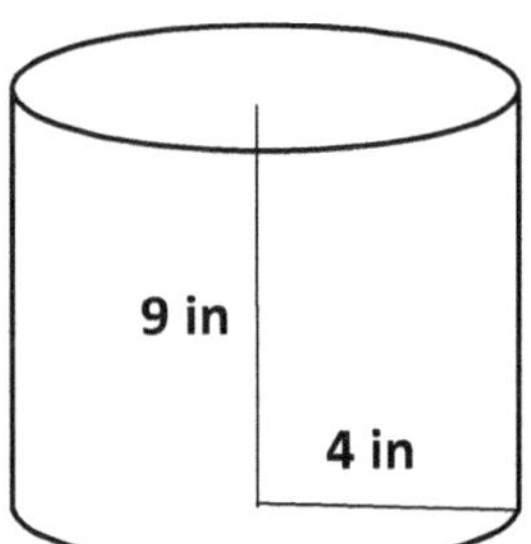

Find the surface area of each Cylinder. ($\pi = 3.14$)

7)

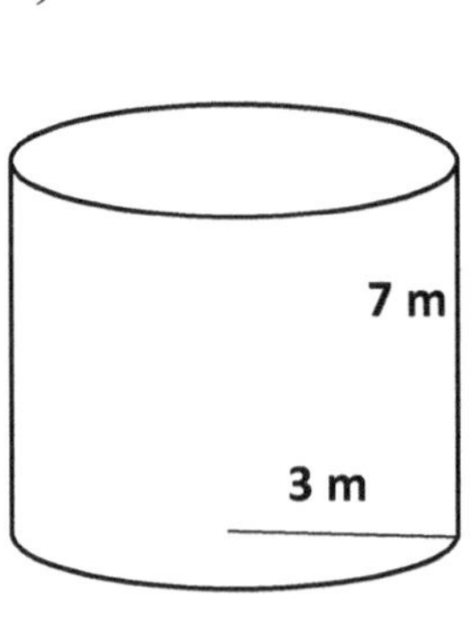

8)

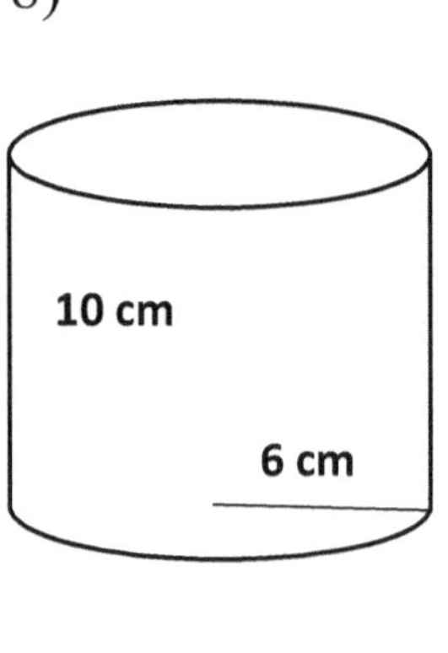

9)

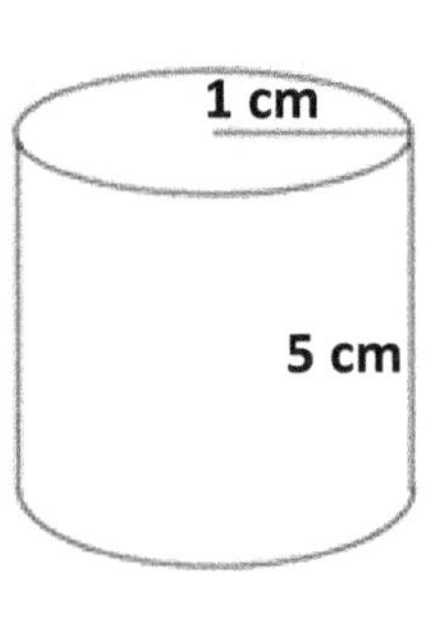

10)

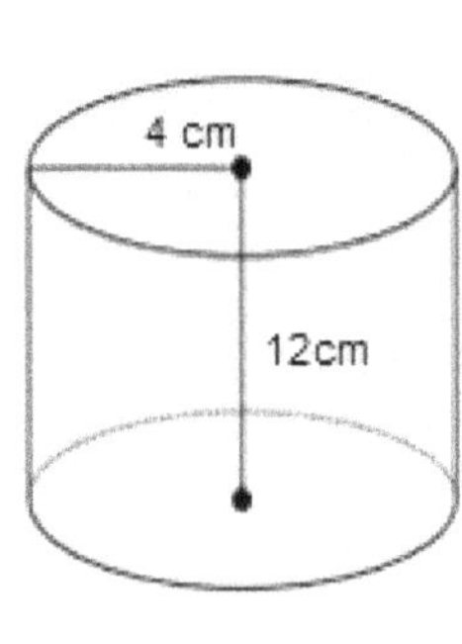

Pyramids and Cone

Find the volume of each Pyramid and Cone. ($\pi = 3.14$)

1)

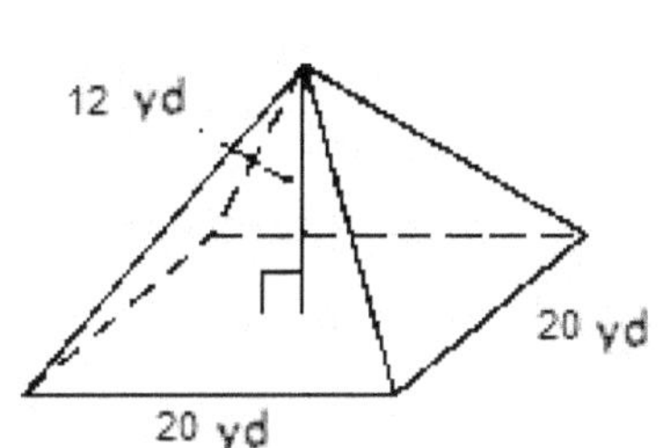

2)

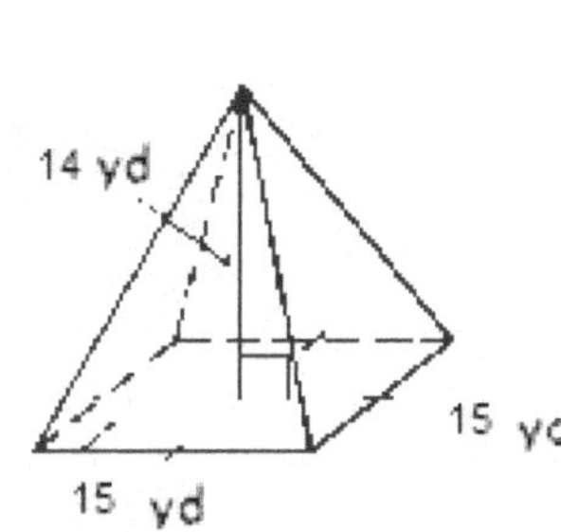

3)

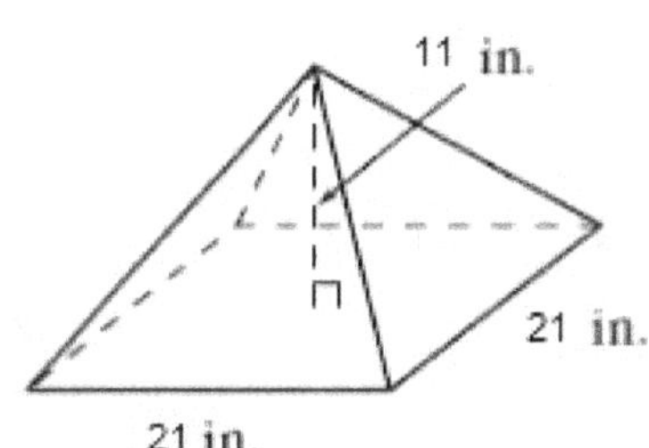

4)

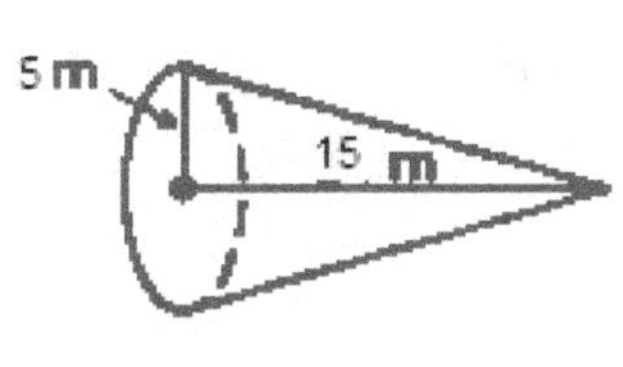

5)

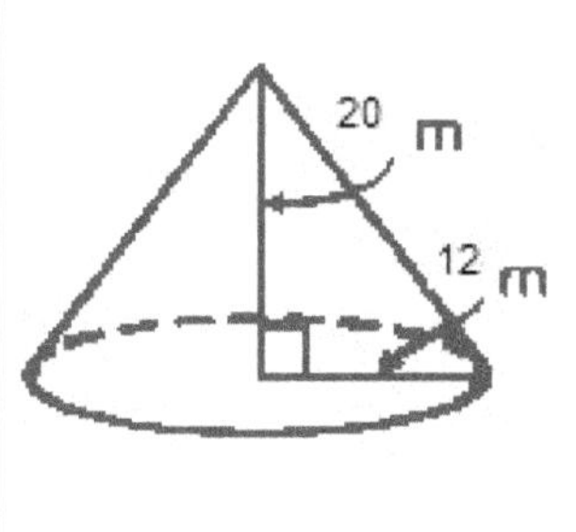

6)

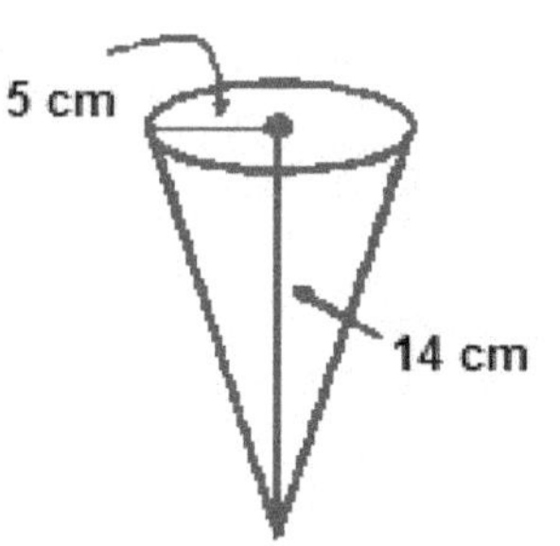

Find the surface area of each Pyramid and Cone. ($\pi = 3.14$)

7)

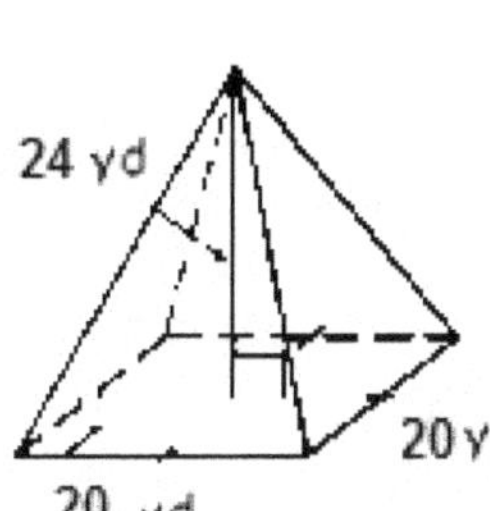

8)

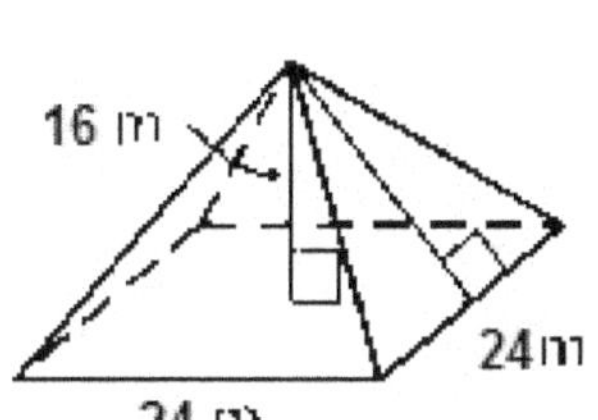

9)

12 in

9 in.

10)

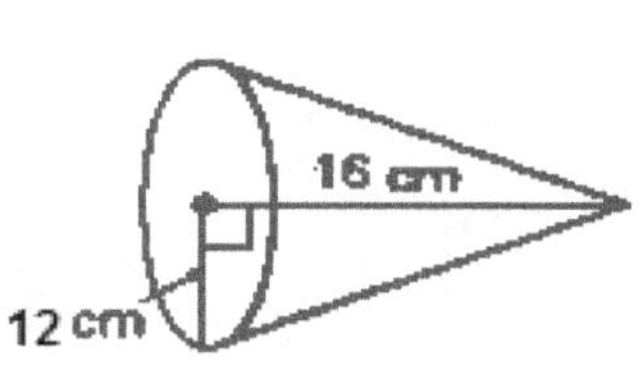

Answers of Worksheets

Angles

1) 16°	4) 34°	7) 90°	10) 75°
2) 96°	5) 70°	8) 75°	11) 70°
3) 59°	6) 52°	9) 33°	

Pythagorean Relationship

1) No	5) Yes	9) 13	13) 15
2) Yes	6) No	10) 20	14) 30
3) No	7) Yes	11) 17	15) 36
4) Yes	8) Yes	12) 10	16) 12

Triangles

1) 60°	5) 45°	9) 54 $square\ unites$
2) 48°	6) 40°	10) 120 $square\ unites$
3) 55°	7) 45°	11) 90 $square\ unites$
4) 52°	8) 67°	12) 180 $square\ unites$

Polygons

1) $52\ ft$	5) $54\ m$	9) $30\ m^2$
2) $26\ in$	6) $25\ cm$	10) $300\ in^2$
3) $112\ ft$	7) $38\ in$	11) $160\ km^2$
4) $20\ cm$	8) $24\ m$	12) $49\ in^2$

Trapezoids

1) $50\ cm^2$	4) $60\ cm^2$	7) 36
2) $105\ m^2$	5) 80	8) 15
3) $39\ ft^2$	6) 24	

Calculate

1) $13\ cm$	2) $11\ ft$	3) $12\ m$	4) $25\ ft$

Circles

1) $19.63\ in^2$	5) $379.94\ cm^2$	9) $12.56\ m^2$
2) $78.5\ cm^2$	6) $314\ miles^2$	10) $113.04\ cm^2$
3) $254.34\ ft^2$	7) $132.67\ in^2$	11) $38.47\ miles^2$
4) $12.56\ m^2$	8) $7.07\ ft^2$	12) $50.24\ ft^2$

Circle No.	Radius	Diameter	Circumference	Area
1	1 in	2 in	6.28 in	3.14 in^2
2	5 m	10 m	31.4 m	78.5 m^2
3	3 ft	6 ft	18.84 ft	28.26 ft^2
4	7.5 miles	15 mi	47.1 mi	176.63 mi^2
5	5.5 km	11 km	34.54 km	94.99 km^2
6	7 cm	14 cm	43.96 cm	153.86 cm^2
7	6 ft	12 ft	37.68 feet	113.04 ft^2
8	10 m	20 m	62.8 m	314 m^2
9	9 in	18 in	56.52 in	254.34 in^2
10	4.5 ft	9 ft	28.26 ft	63.585 ft^2

Cubes

1) 12
2) 8 cm^3
3) 216 ft^3
4) 1,331 m^3
5) 2,197 in^3
6) 343 $miles^3$
7) 1.728 km^3
8) 729 cm^3
9) 9.261 ft^3
10) 1,728 mm^3
11) 0.008 in^3
12) 0.001 km^3
13) 27
14) 294 m^2
15) 150 ft^2
16) 121.5 mm^2
17) 7.26 km^2
18) 726 cm^2

Rectangular Prism

1) 132 m^3
2) 800 in^3
3) 600 m^3
4) 112 cm^3
5) 288 ft^3
6) 420 m^3
7) 184 cm^2
8) 252 ft^2
9) 220 in^2
10) 438 m^2

Cylinder

1) 1,004.8 m^3
2) 214.6 cm^3
3) 9,495.4 cm^3
4) 1.1 m^3
5) 588.8 m^3
6) 452.2 in^3
7) 188.4 m^2
8) 602.9 cm^2
9) 37.7 cm^2
10) 401.9 m^2

Pyramids and Cone

1) 1,600 yd^3
2) 1,050 yd^3
3) 1,617 in^3
4) 392.5 m^3
5) 3,014.4 m^3
6) 366.33 cm^3
7) 1,440 yd^2
8) 1,536 m^2
9) 678.24 in^2
10) 1,205.76 cm^2

Chapter 14 :

Trigonometric Functions

Topics that you'll practice in this chapter:

- ✓ Trig ratios of General Angles
- ✓ Sketch Each Angle in Standard Position
- ✓ Finding Co–Terminal Angles and Reference Angles
- ✓ Angles in Radians
- ✓ Angles in Degrees
- ✓ Evaluating Each Trigonometric Expression
- ✓ Missing Sides and Angles of a Right Triangle
- ✓ Arc Length and Sector Area

Mathematics is like checkers in being suitable for the young, not too difficult, amusing, and without peril to the state. — Plato

Trig ratios of General Angles

Evaluate.

1) $sin\,135° =$ ________

2) $sin\,300° =$ ________

3) $cos - 225° =$ ________

4) $cos\,270° =$ ________

5) $sin\,450° =$ ________

6) $sin\,-330° =$ ________

7) $tan\,60° =$ ________

8) $cot\,180° =$ ________

9) $tan\,240° =$ ________

10) $cot\,90° =$ ________

11) $sec\,180° =$ ________

12) $csc\,90° =$ ________

13) $cot\,-270° =$ ________

14) $sec\,360° =$ ________

15) $cos - 45° =$ ________

16) $sec\,120° =$ ________

17) $csc360° =$ ________

18) $cot\,-45° =$ ________

Find the exact value of each trigonometric function. Some may be undefined.

19) $sec\,2\pi =$ ________

20) $tan - \frac{5\pi}{2} =$ ________

21) $cos\,\frac{11\pi}{2} =$ ________

22) $cot\,\frac{9\pi}{4} =$ ________

23) $sec - 6\pi =$ ________

24) $sec\,\frac{\pi}{4} =$ ________

25) $csc\,\frac{8\pi}{3} =$ ________

26) $cot\,\frac{10\pi}{3} =$ ________

27) $csc - \frac{\pi}{2} =$ ________

28) $cot\,\frac{2\pi}{3} =$ ________

Sketch Each Angle in Standard Position

Draw each angle with the given measure in standard position.

1) $-570°$

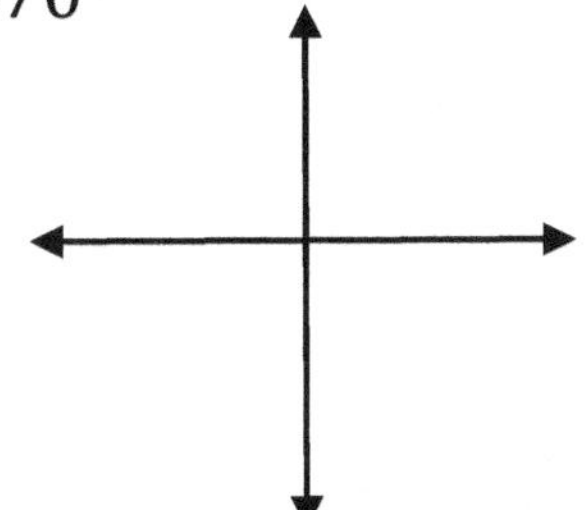

4) $-690°$

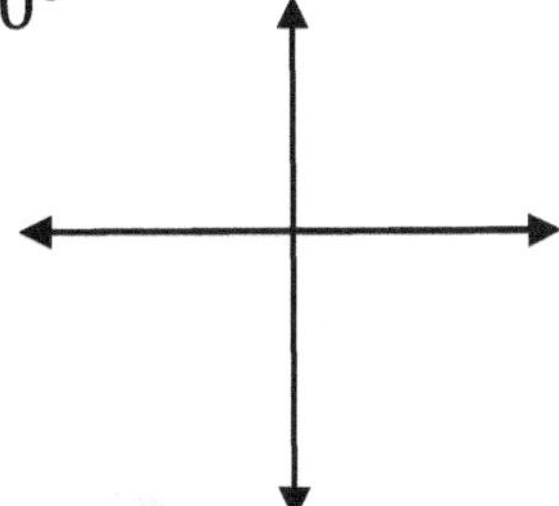

2) $750°$

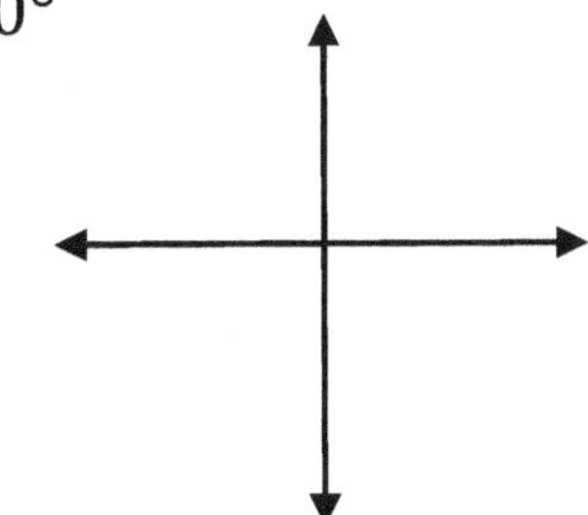

5) $\frac{13\pi}{6}$

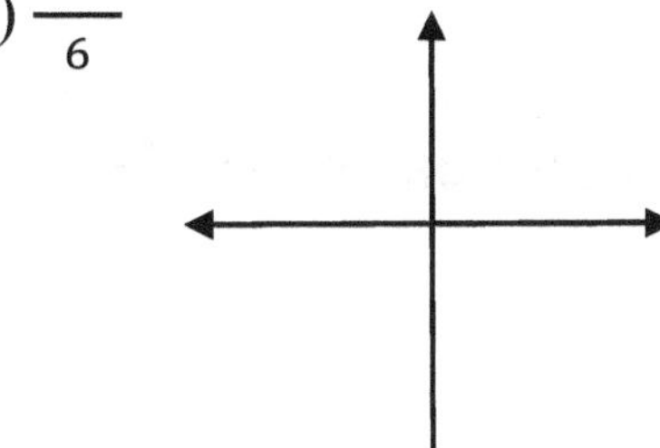

3) $1{,}110°$

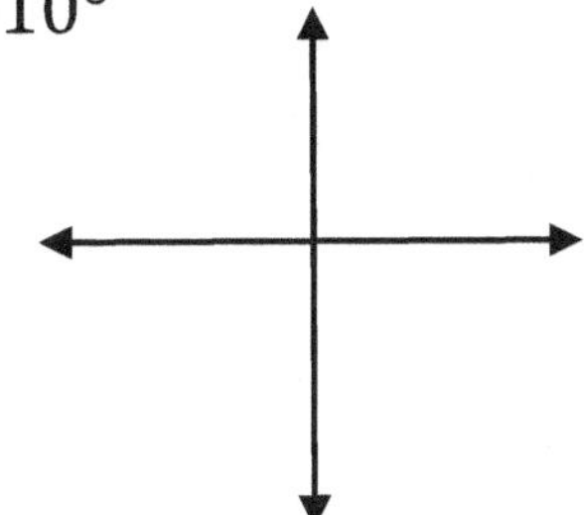

6) $-\frac{11\pi}{6}$

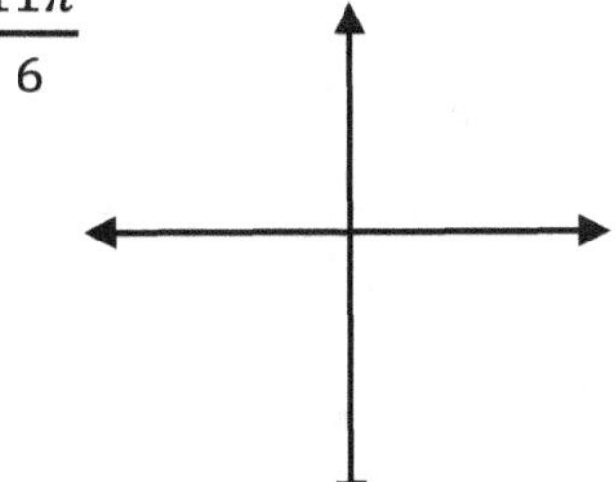

Finding Co-terminal Angles and Reference Angles

Find a conterminal angle between 0° and 360° for each angle provided.

1) $-315° =$

2) $-210° =$

3) $-225° =$

4) $-540° =$

Find a conterminal angle between 0 and 2π for each given angle.

5) $\frac{18\pi}{5} =$

6) $-\frac{19\pi}{6} =$

7) $-\frac{13\pi}{4} =$

8) $\frac{14\pi}{3} =$

Find the reference angle of each angle.

9)

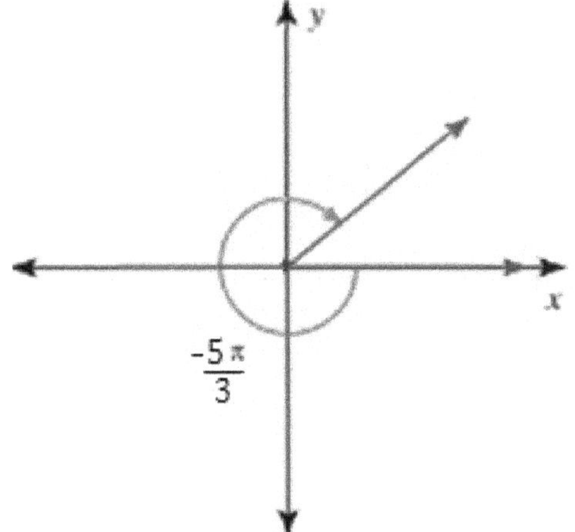

10)

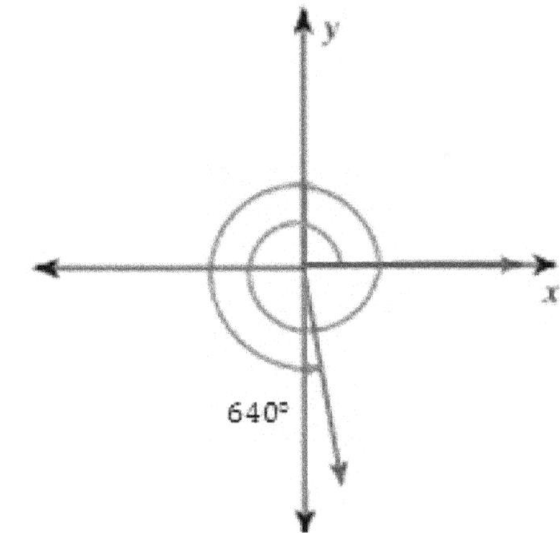

Angles and Angle Measure

Convert each degree measure into radians.

1) $216^\circ =$ ____
2) $660^\circ =$ ____
3) $420^\circ =$ ____
4) $220^\circ =$ ____
5) $210^\circ =$ ____
6) $270^\circ =$ ____
7) $-300^\circ =$ ____
8) $810^\circ =$ ____
9) $330^\circ =$ ____
10) $140^\circ =$ ____
11) $480^\circ =$ ____
12) $405^\circ =$ ____
13) $-450^\circ =$ ____
14) $-126^\circ =$ ____
15) $-675^\circ =$ ____
16) $150^\circ =$ ____
17) $-468^\circ =$ ____
18) $340^\circ =$ ____
19) $-440^\circ =$ ____
20) $342^\circ =$ ____
21) $230^\circ =$ ____

Convert each radian measure into degrees.

22) $\frac{\pi}{10} =$
23) $\frac{5\pi}{12} =$
24) $\frac{7\pi}{3} =$
25) $\frac{3\pi}{20} =$
26) $-\frac{6\pi}{5} =$
27) $\frac{11\pi}{18} =$
28) $-\frac{14\pi}{5} =$
29) $\frac{5\pi}{18} =$
30) $\frac{7\pi}{36} =$
31) $\frac{17\pi}{18} =$
32) $-\frac{13\pi}{30} =$
33) $\frac{7\pi}{9} =$
34) $-\frac{19\pi}{18} =$
35) $\frac{7\pi}{60} =$
36) $-\frac{3\pi}{10} =$
37) $\frac{11\pi}{30} =$
38) $-\frac{2\pi}{9} =$
39) $-\frac{7\pi}{10} =$

Evaluating Trigonometric Functions

Find the exact value of each trigonometric function.

1) $cos\ 780° =$ ________

2) $tan\ \frac{5\pi}{3} =$ ________

3) $\tan -\frac{\pi}{6} =$ ________

4) $\cot -\frac{9\pi}{4} =$ ________

5) $cos\ -\frac{7\pi}{6} =$ ________

6) $\cos 135° =$ ________

7) $sin\ 240° =$ ________

8) $tan\ 330° =$ ________

9) $cot\ 420° =$ ________

10) $tan\ -495° =$ ________

11) $cot\ 315° =$ ________

12) $sin\ -240° =$ ________

13) $cot225° =$ ________

Use the given point on the terminal side of angle θ to find the value of the trigonometric function indicated.

14) $sin\theta;\ (-6, 8)$

15) $cos\theta;\ (-6, 8)$

16) $sec\theta;\ (3,\ 5)$

17) $cos\theta;\ (10, 24)$

18) $sin\theta;\ (6, -6)$

19) $tan\theta;\ (-2, -\sqrt{12})$

Missing Sides and Angles of a Right Triangle

✍ **Find the value of each trigonometric ratio as fractions in their simplest form.**

1) $cot\ x$

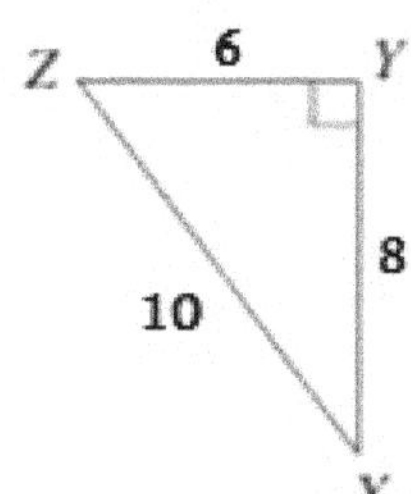

2) $cos\ A$

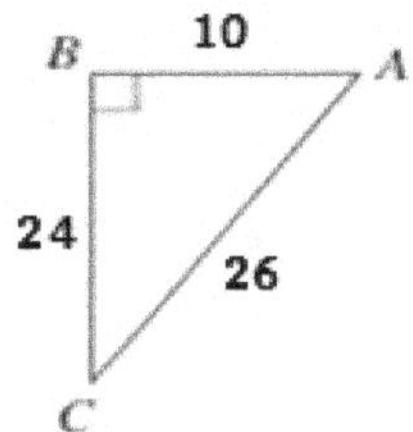

✍ **Find the missing sides. Round answers to the nearest tenth.**

3)

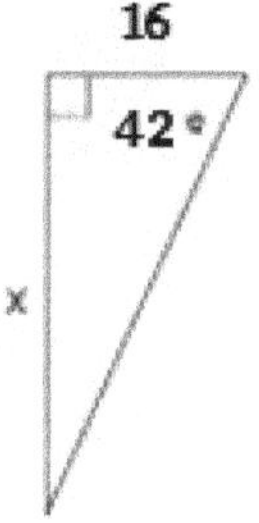

4)

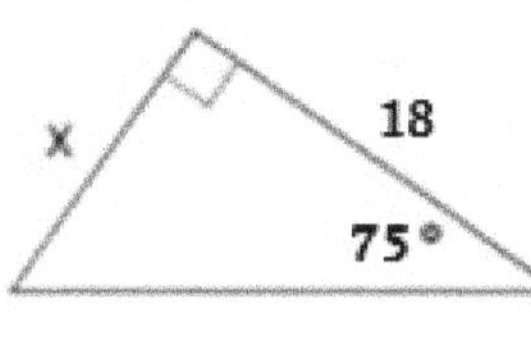

5)

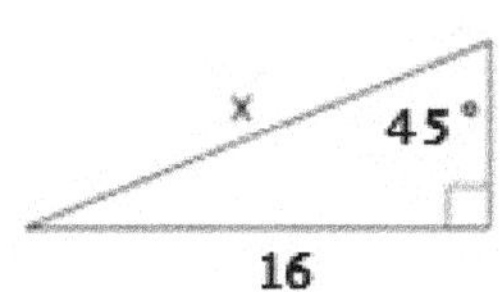

6)

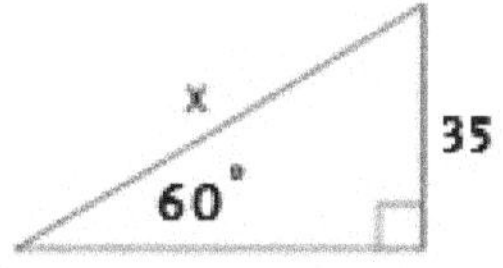

Arc Length and Sector Area

Find the length of each arc. Round your answers to the nearest tenth.

($\pi = 3.14$)

1) $r = 28$ cm, $\theta = 30^\circ$

2) $r = 14$ ft, $\theta = 95^\circ$

3) $r = 22$ ft, $\theta = 50^\circ$

4) $r = 16\,m$, $\theta = 85^\circ$

Find area of each sector. Do *not* round. Round your answers to the nearest tenth. ($\pi = 3.14$)

5)

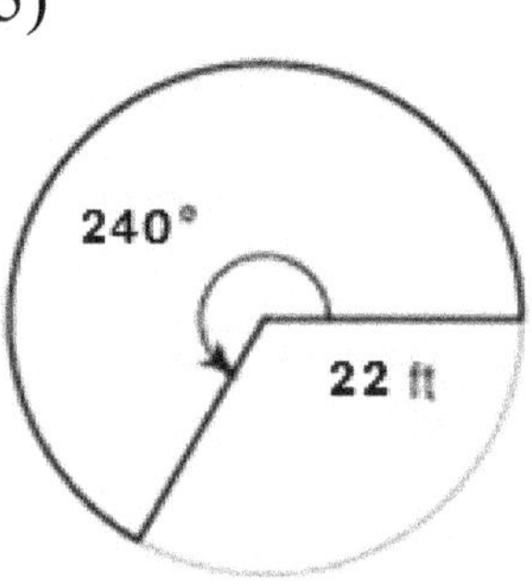

6)

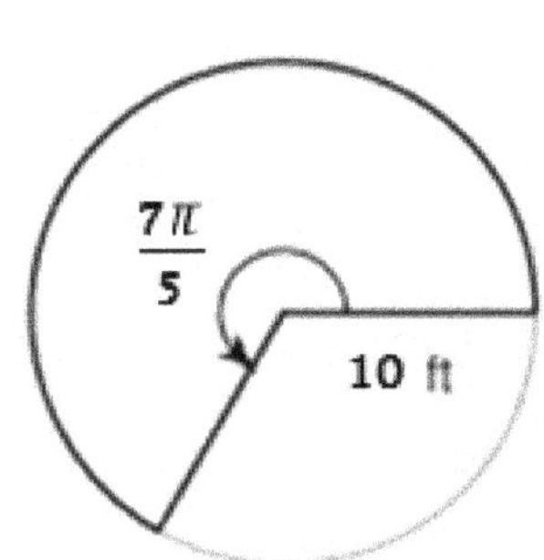

7)

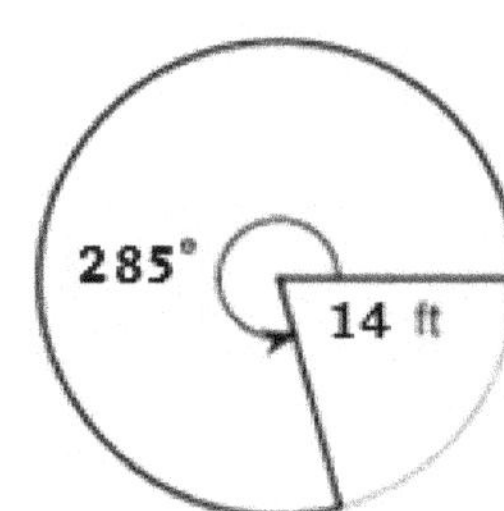

8)

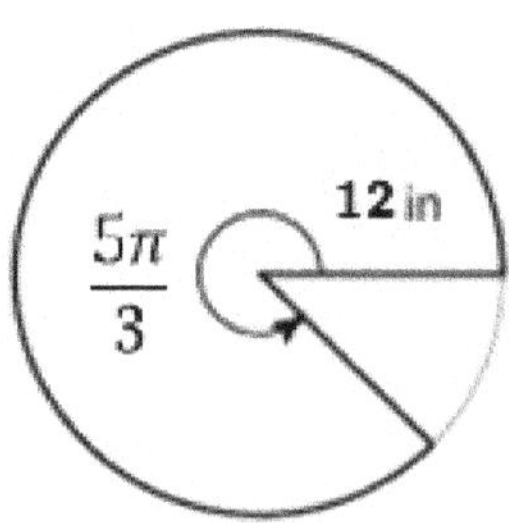

Answers of Worksheets

Trig Ratios of General Angles

1) $\frac{\sqrt{2}}{2}$
2) $-\frac{\sqrt{3}}{2}$
3) $-\frac{\sqrt{2}}{2}$
4) 0
5) 1
6) $\frac{1}{2}$
7) $\sqrt{3}$
8) Undefined
9) $\sqrt{3}$
10) 0
11) -1
12) 1
13) 0
14) 1
15) $\frac{\sqrt{2}}{2}$
16) -2
17) Undefined
18) -1
19) 1
20) Undefined
21) 0
22) 1
23) 1
24) $\sqrt{2}$
25) $\frac{2\sqrt{3}}{3}$
26) $\frac{\sqrt{3}}{3}$
27) -1
28) $-\frac{\sqrt{3}}{3}$

Sketch Each Angle in Standard Position

1) $-570°$

2) $750°$

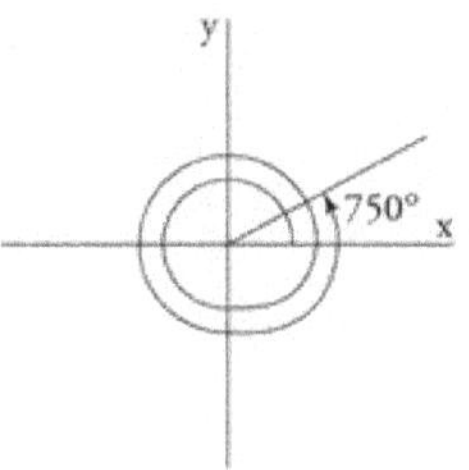

3) $1,110°$

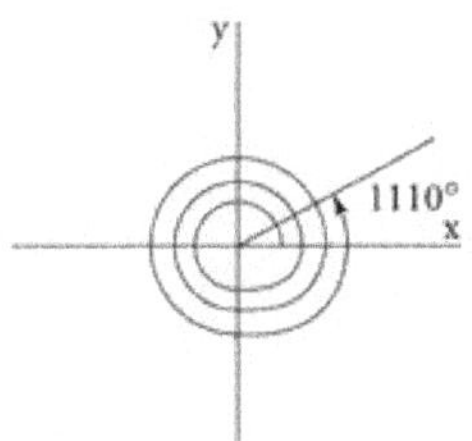

4) $-690°$

5) $\frac{13\pi}{6} = 390°$

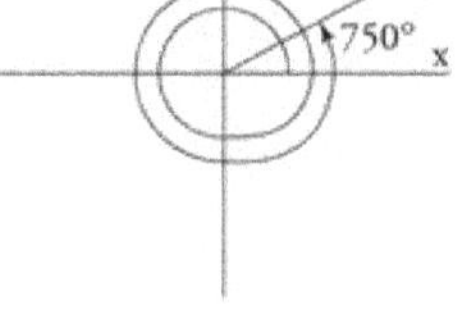

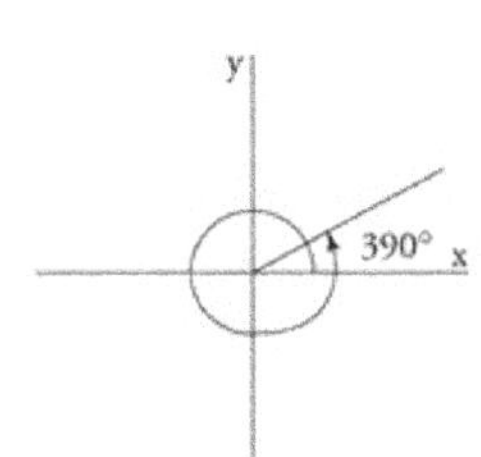

6) $-\frac{11\pi}{6} = -330°$

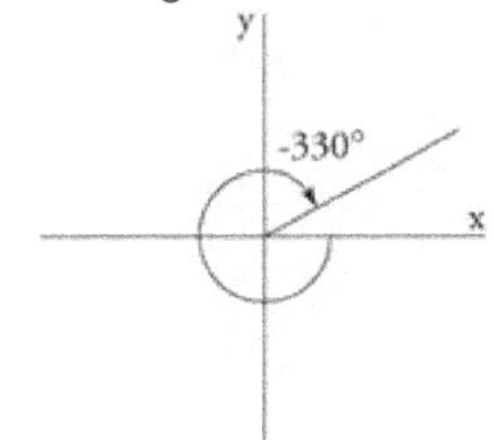

Finding Co–Terminal Angles and Reference Angles

1) $45°$
2) $150°$
3) $135°$
4) $180°$
5) $\frac{4\pi}{5}$
6) $\frac{5\pi}{6}$
7) $\frac{3\pi}{4}$
8) $\frac{2\pi}{3}$
9) $\frac{\pi}{3}$
10) $80°$

Angles and Angle Measure

1) $\frac{6\pi}{5}$
2) $\frac{11\pi}{3}$
3) $\frac{7\pi}{3}$
4) $\frac{11\pi}{9}$

5) $\frac{7\pi}{6}$
6) $\frac{3\pi}{2}$
7) $-\frac{5\pi}{3}$
8) $\frac{9\pi}{2}$
9) $\frac{11\pi}{6}$
10) $\frac{7\pi}{9}$
11) $\frac{8\pi}{3}$
12) $\frac{9\pi}{4}$
13) $-\frac{5}{2}\pi$
14) $-\frac{7\pi}{10}$
15) $-\frac{15\pi}{4}$
16) $\frac{5\pi}{6}$
17) $-\frac{13\pi}{5}$
18) $\frac{17\pi}{9}$
19) $-\frac{22\pi}{9}$
20) $\frac{19\pi}{10}$
21) $\frac{23\pi}{18}$
22) $18°$
23) $75°$
24) $420°$
25) $27°$
26) $-216°$
27) $110°$
28) $-504°$
29) $50°$
30) $35°$
31) $170°$
32) $-78°$
33) $140°$
34) $-190°$
35) $21°$
36) $-54°$
37) $66°$
38) $-40°$
39) $-126°$

Evaluating Each Trigonometric Functions

1) $\frac{1}{2}$
2) $-\sqrt{3}$
3) $-\frac{\sqrt{3}}{3}$
4) -1
5) $-\frac{\sqrt{3}}{2}$
6) $-\frac{\sqrt{2}}{2}$
7) $-\frac{\sqrt{3}}{2}$
8) $-\frac{\sqrt{3}}{3}$
9) $\frac{\sqrt{3}}{3}$
10) 1
11) -1
12) $\frac{\sqrt{3}}{2}$
13) 1
14) 0.8
15) -0.6
16) $\frac{\sqrt{34}}{5}$
17) $\frac{5}{13}$
18) $-\frac{\sqrt{2}}{2}$
19) $\sqrt{3}$

Missing Sides and Angles of a Right Triangle

1) $\frac{4}{3}$
2) $\frac{5}{13}$
3) 14.4
4) 67.2
5) 22.6
6) 40.4

Arc Length and Sector Area

1) 14.7 cm
2) 23.2 ft
3) 19.2 ft
4) 23.7m
5) 1,013.7 ft^2
6) 220 in^2
7) 487.5 ft^2
8) 377 in^2

Chapter 15 :

Statistics and Probability

Topics that you'll practice in this chapter:

- ✓ Mean and Median
- ✓ Mode and Range
- ✓ Histograms
- ✓ Stem–and–Leaf Plot
- ✓ Pie Graph
- ✓ Probability Problems
- ✓ Factorials
- ✓ Combinations and Permutation

"The book of nature is written in the language of Mathematic."

- Galileo.

Mean and Median

Find Mean and Median of the Given Data.

1) 8, 7, 14, 4, 8

2) 14, 8, 25, 19, 16, 33, 11

3) 23, 18, 15, 12, 17

4) 34, 14, 10, 15, 6, 11

5) 10, 19, 6, 8, 32, 20, 17

6) 17, 26, 39, 69, 20, 6

7) 40, 38, 18, 11, 9, 2, 7, 32,41

8) 24, 21, 31,12,33, 32, 22

9) 16, 14, 20, 41, 15, 20, 38, 4

10)20, 20, 30, 18, 6, 28, 12, 46

11) 12, 7, 10, 11, 16, 22

12) 10, 29, 27, 12, 2, 15, 10, 3

Calculate.

13)In a javelin throw competition, five athletics score 56, 34, 62, 23 and 19 meters. What are their Mean and Median? ________________

14)Eva went to shop and bought 8 apples, 14 peaches, 6 bananas, 4 pineapples and 12 melons. What are the Mean and Median of her purchase? ________________

15)Bob has 17 black pen, 19 red pen, 14 green pens, 20 blue pens and 5 boxes of yellow pens. If the Mean and Median are 19 respectively, what is the number of yellow pens in each box? ________________

Mode and Range

Find Mode and Rage of the Given Data.

1) 4, 3, 7, 3, 3, 4

Mode: _____ Range: _____

2) 18, 18, 24, 26, 18, 8, 14, 22

Mode: _____ Range: _____

3) 8, 8, 8, 16, 19, 22, 20, 9, 13

Mode: _____ Range: _____

4) 24, 24, 14, 28, 20, 18, 20, 24

Mode: _____ Range: _____

5) 6, 21, 27, 24, 27, 27

Mode: _____ Range: _____

6) 21, 8, 8, 7, 8, 12, 10, 22, 18, 13

Mode: _____ Range: _____

7) 7, 4, 4, 6, 13, 13, 13, 0, 2, 2

Mode: _____ Range: _____

8) 5, 8, 5, 14, 12, 14, 3, 5, 18

Mode: _____ Range: _____

9) 7, 7, 7, 12, 7, 3, 8, 16, 3, 17

Mode: _____ Range: _____

10) 15, 15, 19, 16, 4, 16, 10, 15

Mode: _____ Range: _____

11) 6, 6, 5, 6, 42, 13, 19, 2

Mode: _____ Range: _____

12) 8, 8, 9, 8, 9, 4, 34, 22

Mode: _____ Range: _____

Calculate.

13) A stationery sold 12 pencils,56 red pens,24 blue pens,20 notebooks, 12 erasers, 21 rulers and 11 color pencils. What are the Mode and Range for the stationery sells?

Mode: _____ Range: _____

14) In an English test, eight students score 10, 15, 15, 18 18, 16, 15 and 15. What are their Mode and Range? _______________

15) What is the range of the first 6 even numbers greater than 8?

____

Times Series

Use the following Graph to complete the table.

Day	Distance (km)
1	
2	

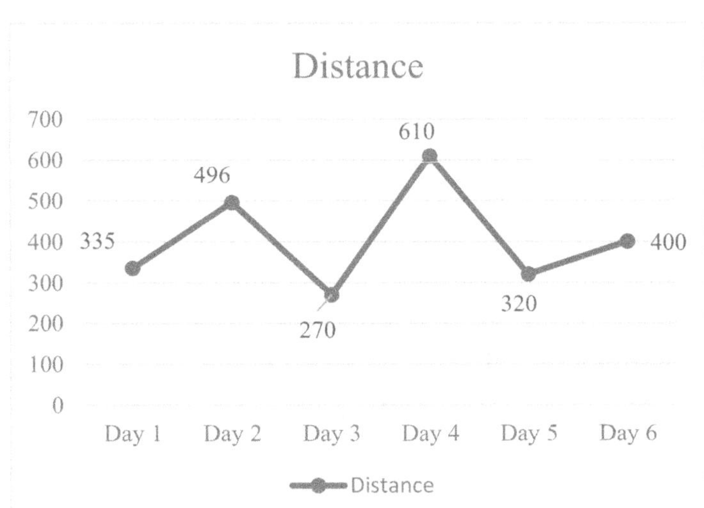

The following table shows the number of births in the US from 2007 to 2012 (in millions).

Year	Number of births (in millions)
2007	4.15
2008	3.70
2009	3.45
2010	3.20
2011	1.75
2012	2.98

Draw a Time Series for the table.

Stem–and–Leaf Plot

Make stem ad leaf plots for the given data.

1) 24, 26,29, 20, 53, 27, 51, 55, 36, 21, 37, 30

Stem	Leaf plot

2) 11, 59, 66, 14, 18, 19, 59, 65, 69, 61, 68, 65

Stem	Leaf plot

3) 121, 55, 66, 54, 112, 128, 63, 125, 59, 123, 68, 119

Stem	Leaf plot

4) 51, 32, 100, 56, 84, 36, 107, 56, 85, 39, 56, 106, 89

Stem	Leaf plot

5) 33, 89, 19, 87, 81, 16, 11, 30, 86, 35, 17, 35, 13

Stem	Leaf plot

6) 60, 92, 22, 25, 67, 93, 95, 62, 21, 64, 98, 29

Stem	Leaf plot

Pie Graph

The circle graph below shows all Robert's expenses for last month. Robert spent $140 on his hobbies last month.

Answer following questions based on the Pie graph.

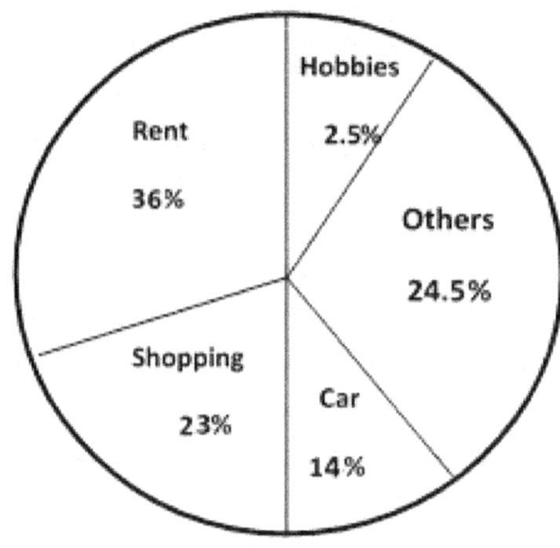

1) How much was Robert's total expenses last month? ________________

2) How much did Robert spend on his car last month? ________________

3) How much did Robert spend for shopping last month? __________

4) How much did Robert spend on his rent last month? ________________

5) What fraction is Robert's expenses for his rent and car out of his total expenses last month? __________________

Probability Problems

Calculate.

1) A number is chosen at random from 1 to 10. Find the probability of selecting number 6 or smaller numbers. ________________

2) Bag A contains 18 red marbles and 6 green marbles. Bag B contains 16 black marbles and 8 orange marbles. What is the probability of selecting a green marble at random from bag A? What is the probability of selecting a black marble at random from Bag B? ________________

3) A number is chosen at random from 1 to 20. What is the probability of selecting multiples of 4? ________________

4) A card is chosen from a well-shuffled deck of 52 cards. What is the probability that the card will be a queen? ________________

5) A number is chosen at random from 1 to 15. What is the probability of selecting a multiple of 3 or 5? ________________

A spinner numbered 1–8, is spun once. What is the probability of spinning ...?

6) an Odd number? __________ 7) a multiple of 2? _____

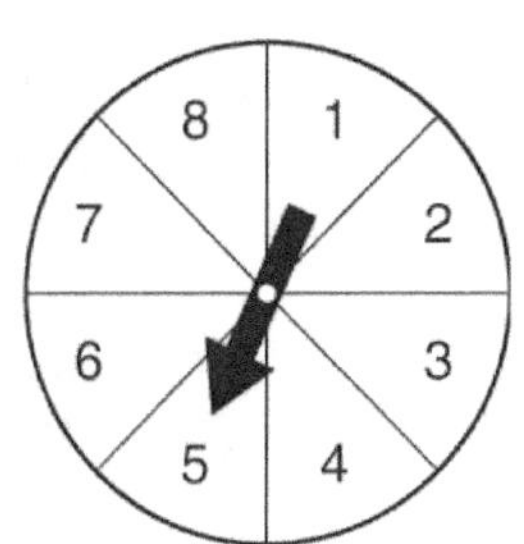

8) a multiple of 5? _____ 9) number 10? ________

Factorials

Determine the value for each expression.

1) $4!\ +0!\ =$

2) $2!+5!\ =$

3) $(2!)^2 =$

4) $5!-3! =$

5) $6!-3!+10 =$

6) $3!\times 4-15 =$

7) $(2!+3!)^2 =$

8) $(4!-3!)^2 =$

9) $(3!\,0!)^2-10 =$

10) $\frac{10!}{8!} =$

11) $\frac{6!}{4!} =$

12) $\frac{6!}{5!} =$

13) $\frac{15!}{13!} =$

14) $\frac{n!}{(n-3)!} =$

15) $\frac{(n+2)!}{n!} =$

16) $\frac{(2+2!)^3}{2!} =$

17) $\frac{5(n+2)!}{(n+1)!} =$

18) $\frac{22!}{20!4!} =$

19) $\frac{13!}{11!3!} =$

20) $\frac{9\times 210!}{3(7\times 30)!} =$

21) $\frac{32!}{31!2!} =$

22) $\frac{11!12!}{10!13!} =$

23) $\frac{16!15!}{14!14!} =$

24) $\frac{(5\times 3)!}{0!14!} =$

25) $\frac{4!(5n-2)!}{(5n)!} =$

26) $\frac{4n(4n+7)!}{(4n+8)!} =$

27) $\frac{(n-2)!(n+1)}{(n+2)!} =$

Combinations and Permutations

Calculate the value of each.

1) $6! =$ ____
2) $2! \times 5! =$ ____
3) $3 \times 4! =$ ____
4) $5! + 3! =$ ____
5) $7! =$ ____
6) $4! =$ ____
7) $3! + 3! =$ ____
8) $7! - 5! =$ ____

Find the answer for each word problems.

9) Susan is baking cookies. She uses sugar, butter, Vanilla, eggs and flour. How many different orders of ingredients can she try? ____________

10) Albert is planning for his vacation. He wants to go to museum, watch a movie, go to the beach, play the game and play football. How many ways of ordering are there for him? ____________

11) How many 4-digit numbers can be named using the digits 3, 4, 5, and 6 without repetition? ____________

12) In how many ways can 5 boys be arranged in a straight line? ____________

13) In how many ways can 6 athletes be arranged in a straight line? ____________

14) A professor is going to arrange her 7 students in a straight line. In how many ways can she do this? ____________

15) How many code symbols can be formed with the letters for the word GAMES? ____________

16) In how many ways a team of 7 basketball players can choose a captain and co-captain? ____________

Answers of Worksheets

Mean and Median

1) Mean: 8.2, Median: 8
2) Mean: 18, Median: 16
3) Mean: 17, Median: 17
4) Mean: 15, Median: 12.5
5) Mean: 16, Median: 17
6) Mean: 29.5, Median: 23
7) Mean: 22, Median: 18
8) Mean: 25, Median: 24
9) Mean: 21, Median: 18
10) Mean: 22.5, Median: 20
11) Mean: 13, Median: 11.5
12) Mean: 13.5, Median: 11
13) Mean: 38.8, Median: 34
14) Mean: 8.8, Median: 8
15) 5

Mode and Range

1) Mode: 3, Range: 4
2) Mode: 18, Range: 18
3) Mode: 8, Range: 14
4) Mode: 24, Range: 14
5) Mode: 27, Range: 21
6) Mode: 8, Range: 15
7) Mode: 13, Range: 13
8) Mode: 5, Range: 15
9) Mode: 7, Range: 14
10) Mode: 15, Range: 15
11) Mode: 6, Range: 40
12) Mode: 8, Range: 30
13) Mode: 12, Range: 45
14) Mode: 15, Range: 8
15) 10

Time series

Day	Distance (km)
1	335
2	496
3	270
4	610
5	320
6	400

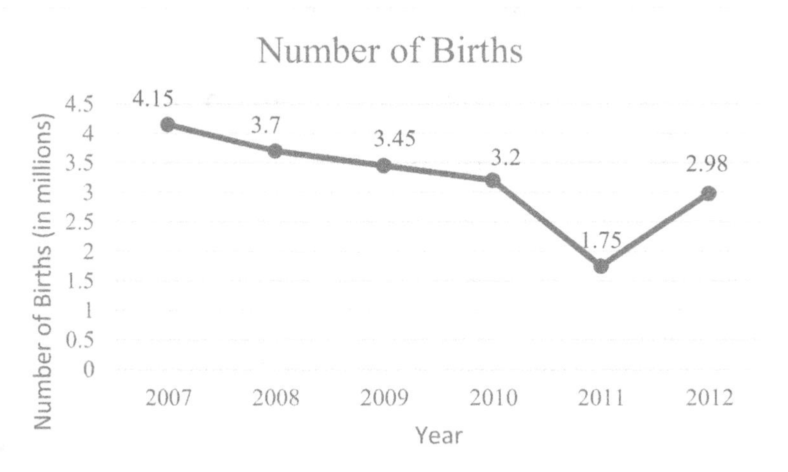

Stem–And–Leaf Plot

1)

Stem	leaf
2	0 1 4 6 7 9
3	0 6 7
5	1 3 5

2)

Stem	leaf
1	1 4 8 9
5	9 9
6	1 5 5 6 8 9

3)

Stem	leaf
5	4 5 9
6	3 6 8
11	2 9
12	1 3 5 8

4)

Stem	leaf
3	2 6 9
5	1 6 6 6
8	4 5 9
10	0 6 7

5)

Stem	leaf
1	1 3 6 7 9
3	0 3 5 5
8	1 6 7 9

6)

Stem	leaf
2	2 1 5 9
6	0 2 4 7
9	2 3 5 8

Pie Graph

1) \$5,600
2) \$784
3) \$1,288
4) \$2,016
5) $\frac{1}{2}$

Probability Problems

1) $\frac{3}{5}$
2) $\frac{1}{4}, \frac{2}{3}$
3) $\frac{1}{4}$
4) $\frac{1}{13}$
5) $\frac{7}{15}$
6) $\frac{1}{2}$
7) $\frac{1}{2}$
8) $\frac{1}{8}$
9) 0

Factorials

1) 25
2) 122
3) 4
4) 114
5) 724
6) 9
7) 64
8) 324
9) 26
10) 90
11) 30
12) 6
13) 210
14) $n(n-1)(n-2)$
15) $(n+1)(n+2)$
16) 32
17) $5(n+2)$
18) 19.25
19) 26
20) 3
21) 16
22) $\frac{11}{13}$
23) 3,600
24) 15
25) $\frac{24}{5n(5n-1)}$
26) $\frac{n}{(n+2)}$
27) $\frac{1}{n(n-1)(n+2)}$

Combinations and Permutations

1) 720
2) 240
3) 72
4) 126
5) 5,040
6) 24
7) 12
8) 4,920
9) 120
10) 120
11) 24
12) 120
13) 720
14) 5,040
15) 120
16) 42

Chapter 16 :

SAT Math Test Review

The SAT is a standardized test widely used for college admissions in the United States. It is a pencil-and-paper test developed and administered by the College Board, a private and non-profit organization.

The SAT test is divided into three major segments.

- Mathematics (two sections)
- Reading
- Writing

There are two Math sections on the SAT test:

✓ **Math Test Section 1 (No Calculator):** this section has 20 questions (15 multiple choice and 5 grid-in) and lasts 25 minutes.

✓ **Math Test Section 2 (Calculator):** This section has 38 questions (30 multiple choice and 8 grid-in) and lasts 55 minutes.

The SAT Math Test covers a range of math practices including the heart of algebra, passport to advanced math, problem solving & data analysis. In total, the SAT math test is 80 minutes long and includes 58 questions: 45 multiple choice questions and 13 grid-in questions. In a SAT assessment test, all questions are weighted the same and there is no penalty for wrong answers in the SAT test.

In this section, there are two complete SAT Math Tests. Take these tests to see what score you'll be able to receive on a real SAT Math test.

The hardest arithmetic to master is that which enables us to count our blessings.

~Eric Hoffer

Time to Test

Time to refine your skill with a practice examination.

Take a practice SAT Math Test to simulate the test day experience. After you've finished, score your test using the answer key.

Before You Start

- You'll need a pencil, a calculator and a timer to take the test.
- For each question, there are four possible answers. Choose which one is best.
- It's okay to guess. There is no penalty for wrong answers.
- Use the answer sheet provided to record your answers.
- After you've finished the test, review the answer key to see where you went wrong.

Good Luck!

SAT Math Practice Test Answer Sheets

Remove (or photocopy) these answer sheets and use them to complete the practice tests.

SAT Practice Test Section 1

1	Ⓐ Ⓑ Ⓒ Ⓓ	6	Ⓐ Ⓑ Ⓒ Ⓓ	11	Ⓐ Ⓑ Ⓒ Ⓓ
2	Ⓐ Ⓑ Ⓒ Ⓓ	7	Ⓐ Ⓑ Ⓒ Ⓓ	12	Ⓐ Ⓑ Ⓒ Ⓓ
3	Ⓐ Ⓑ Ⓒ Ⓓ	8	Ⓐ Ⓑ Ⓒ Ⓓ	13	Ⓐ Ⓑ Ⓒ Ⓓ
4	Ⓐ Ⓑ Ⓒ Ⓓ	9	Ⓐ Ⓑ Ⓒ Ⓓ	14	Ⓐ Ⓑ Ⓒ Ⓓ
5	Ⓐ Ⓑ Ⓒ Ⓓ	10	Ⓐ Ⓑ Ⓒ Ⓓ	15	Ⓐ Ⓑ Ⓒ Ⓓ

16

17

18

19

20

SAT Practice Test Section 2

1	Ⓐ Ⓑ Ⓒ Ⓓ	9	Ⓐ Ⓑ Ⓒ Ⓓ	17	Ⓐ Ⓑ Ⓒ Ⓓ	25	Ⓐ Ⓑ Ⓒ Ⓓ
2	Ⓐ Ⓑ Ⓒ Ⓓ	10	Ⓐ Ⓑ Ⓒ Ⓓ	18	Ⓐ Ⓑ Ⓒ Ⓓ	26	Ⓐ Ⓑ Ⓒ Ⓓ
3	Ⓐ Ⓑ Ⓒ Ⓓ	11	Ⓐ Ⓑ Ⓒ Ⓓ	19	Ⓐ Ⓑ Ⓒ Ⓓ	27	Ⓐ Ⓑ Ⓒ Ⓓ
4	Ⓐ Ⓑ Ⓒ Ⓓ	12	Ⓐ Ⓑ Ⓒ Ⓓ	20	Ⓐ Ⓑ Ⓒ Ⓓ	28	Ⓐ Ⓑ Ⓒ Ⓓ
5	Ⓐ Ⓑ Ⓒ Ⓓ	13	Ⓐ Ⓑ Ⓒ Ⓓ	21	Ⓐ Ⓑ Ⓒ Ⓓ	29	Ⓐ Ⓑ Ⓒ Ⓓ
6	Ⓐ Ⓑ Ⓒ Ⓓ	14	Ⓐ Ⓑ Ⓒ Ⓓ	22	Ⓐ Ⓑ Ⓒ Ⓓ	30	Ⓐ Ⓑ Ⓒ Ⓓ
7	Ⓐ Ⓑ Ⓒ Ⓓ	15	Ⓐ Ⓑ Ⓒ Ⓓ	23	Ⓐ Ⓑ Ⓒ Ⓓ		
8	Ⓐ Ⓑ Ⓒ Ⓓ	16	Ⓐ Ⓑ Ⓒ Ⓓ	24	Ⓐ Ⓑ Ⓒ Ⓓ		

31 32 33 34

35 36 37 38

SAT Math Practice Test 1

Section 1

- ❖ **20 Questions.**
- ❖ **Total time for this test: 25 Minutes.**
- ❖ **You may NOT use a calculator on this Section.**

Administered *Month Year*

1) If $f(x) = 4x + 3(2x + 5) - 7$ then $f(2x) = ?$

A. $20x + 8$

B. $20x - 8$

C. $20x - 7$

D. $20x + 7$

2) A line in the xy-plane passes through origin and has a slope of $\frac{1}{6}$. Which of the following points lies on the line?

A. $(16, 4)$

B. $(24, 6)$

C. $(18, 3)$

D. $(18, 6)$

3) If $x + 2y = 0$, $5x - 3y = 16$, which of the following ordered pairs (x, y) satisfies both equations?

A. $(-1, 3)$

B. $(3, -4)$

C. $(2, -1)$

D. $(-6, -4)$

4) If $6x - 12 = 24$, what is the value of $8x - 20$?

A. 24

B. 36

C. 20

D. 28

5) Which of the following is equivalent to $(12n^2 + 4n + 6) - (7n^2 - 2)$?

A. $5n + 4n^2 + 8$

B. $3n^2 - 7$

C. $5n^2 + 4n + 8$

D. $4n + 8$

6) If $(ax + 7)(bx + 5) = 27x^2 + cx + 35$ for all values of x and $a + b = 12$, what are the two possible values for c?

A. 66, 55

B. 78, 55

C. 66, 78

D. 56, 73

7) If $x \neq -6$ and $x \neq 5$, which of the following is equivalent to $\frac{1}{\frac{1}{x-5}+\frac{1}{x+6}}$?

A. $\frac{(x-5)(x+6)}{(x-5)+(x+6)}$

C. $\frac{(x+6)(x-5)}{(x+6)-(x+5)}$

B. $\frac{(x+6)+(x-5)}{(x+6)(x-5)}$

D. $\frac{(x+6)+(x-5)}{(x+6)-(x-5)}$

8) In the xy-plane, if (0, 0) is a solution to the system of inequalities below, which of the following relationships between a and b must be true?

$$y > b + x \, , y < a + x$$

A. $b > a$

C. $a = b$

B. $b < a$

D. $b = b + a$

9) John buys a pepper plant that is 3 inches tall. With regular watering the plant grows 8 inches a year. Writing John's plant's height as a function of time, what does the y −intercept represent?

A. The y −intercept represents the rate of grows of the plant which is 8 inches.

B. The y −intercept represents the starting height of 3 inches.

C. The y −intercept represents the rate of growth of plant which is 3 inches per year.

D. There is no y −intercept.

10) Which of the following points lies on the line that goes through the points $(2,-2)$ and $(4,-8)$?

A. $(-1,5)$ C. $(2,6)$

B. $(1,3)$ D. $(0,4)$

11) Calculate f(3) for the following function f.

$$f(x) = 3x^3 - 4x + 5$$

A. 47 C. −74

B. 74 D. 64

12) If $\frac{3}{x} = \frac{18}{x-15}$ what is the value of $\frac{x}{3}$?

A. 1.25 C. −1

B. −3.5 D. −1.25

13) What is the equation of the graph?

A. $x^2 + 3x + 10$

B. $3x^2 - x + 10$

C. $x^2 - 6x - 10$

D. $x^2 + +6x + 10$

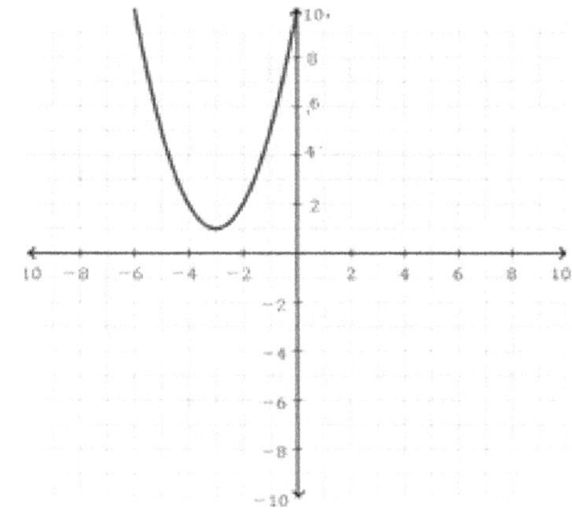

14) Which of the following is an equation of a circle in the xy-plane with center $(3, 5)$ and a radius with endpoint $(\frac{5}{2}, 8)$?

A. $(x+3)^2 + (y-5)^2 = \frac{37}{4}$

B. $x^2 + (y+5)^2 = \frac{37}{4}$

C. $(x+3)^2 + (y+5)^2 = \frac{37}{4}$

D. $(x-3)^2 + (y-5)^2 = \frac{37}{4}$

15) Given a right triangle ΔABC whose $\angle B = 90^\circ$, $\sin C = \frac{3}{5}$, find $\cos A$?

A. 0

B. $\frac{3}{5}$

C. $\frac{7}{3}$

D. $\frac{5}{3}$

Grid-ins Questions

Questions 16–20 are grid-ins questions. Solve the problems and enter your answers in the grid on the answer sheet as shown below.

16) If $x^2 + 8x + r$ factors into $(x + 6)(x + p)$,and r and p are constants, what is the value of r?

17) Sara opened a bank account that earns 12 percent compounded annually. Her initial deposit was \$140, and she uses the expression $\$140(x)^n$ to find the value of the account after n years. What is the value of x in the expression?

18) 7 liters of water are poured into an aquarium that's 70 cm long, 20 cm wide, and 50 cm high. How many centimeters will the water level in the aquarium rise due to this added water? (1 liter of water = 1,000 cm^3)

19) For what real value of x is the equation below true?

$$x^3 - 6x^2 + 3x - 18 = 0$$

20) The length of a rectangle is 7 meters greater than 3 times its width. The perimeter of the rectangle is 78 meters. What is the area of the rectangle in meters?

STOP

This is the End of this Section. You may check your work on this section if you still have time.

SAT Math Practice Test 1

Section 2

- 38 Questions.
- Total time for this test: 55 Minutes.
- You may NOT use a calculator on this Section.

Administered Month Year

1) If $y = nx + 6$, where n is a constant, and when $x = 4$, $y = 14$, what is the value of y when $x = 5$?

A. 32

C. 14

B. 12

D. 16

2) Which of the following numbers is NOT a solution of the inequality $4x - 9 \geq 5x - 3$?

A. −5

C. −10

B. −8

D. −9

3) If the following equations are true, what is the value of x?

$$a = \sqrt{2}$$

$$6a = \sqrt{6x}$$

A. 20

C. 12

B. 18

D. 24

4) If $\sqrt{9m - 20} = m$, what is (are) the value(s) of m?

A. 4

C. 4, 5

B. 9

D. −4, −5

5) If $4n - 13 \geq 2$, what is the least possible value of $4n + 5$?

A. 15

C. 20

B. 2

D. 18

6) What is the ratio of the minimum value to the maximum value of the following function?

$$f(x) = -5x + 3; \qquad -1 \le x \le 3$$

A. $\frac{3}{2}$

C. $-\frac{12}{7}$

B. $-\frac{7}{12}$

D. $\frac{7}{12}$

7) The equation $x^2 = 6x - 9$ has how many distinct real solutions?

A. 2

C. 0

B. No Solution

D. 1

Gender	Under 45	45 or older	total
Male	12	14	26
Female	16	8	24
Total	28	22	50

8) The table above shows the distribution of age and gender for 40 employees in a company. If one employee is selected at random, what is the probability that the employee selected be either a female under age 45 or a male age 45 or older?

A. $\frac{2}{5}$

C. $\frac{3}{25}$

B. $\frac{4}{25}$

D. $\frac{3}{5}$

9) If $\frac{8}{3}y = \frac{16}{5}$, what is the value of y?

A. $\frac{3}{8}$

B. $\frac{5}{6}$

C. $\frac{8}{5}$

D. $\frac{6}{5}$

10) In the triangle below, if the measure of angle A is 30 degrees, then what is the value of y? (figure is NOT drawn to scale)

A. 80

B. 98

C. 88

D. 82

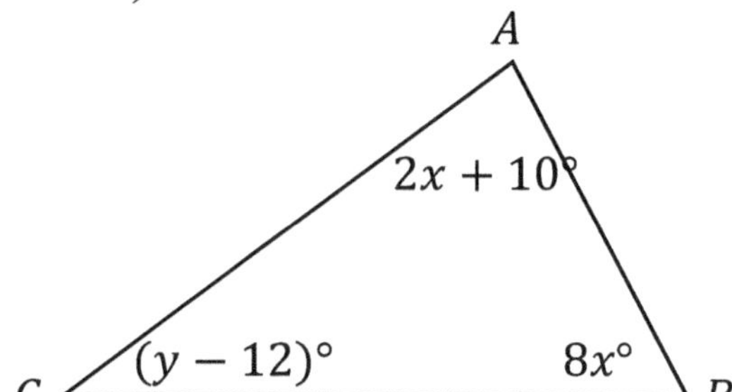

11) If the function g(x) has three distinct zeros, which of the following could represent the graph of g(x)?

A.

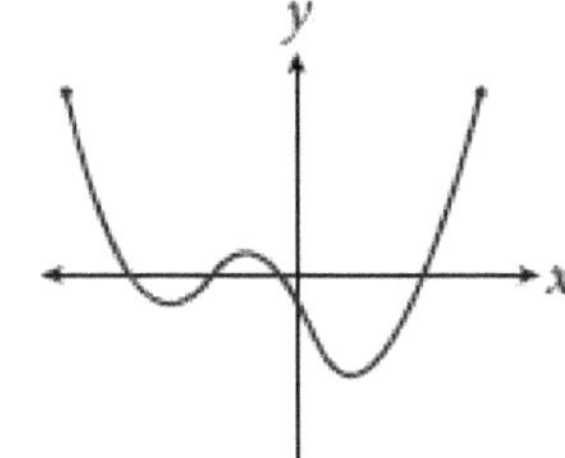

B.

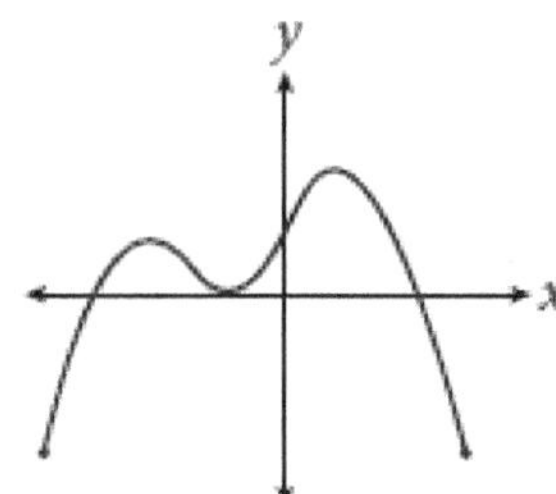

C.

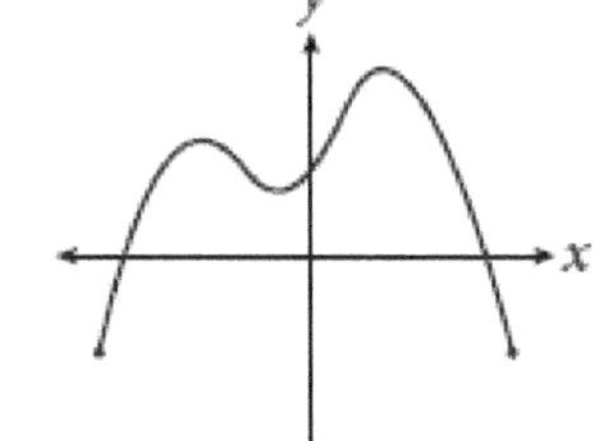

D.

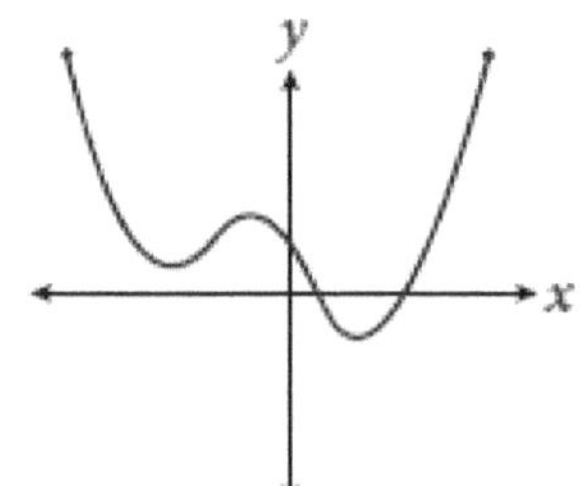

12) If the positive integer x leaves a remainder of 5 when divided by 6, what will the remainder be when $x + 8$ is divided by 6?

A. 3 C. 1

B. 0 D. 4

13) Jack walks 50 meters in 40 seconds. If he walks at this same rate, which of the following is the distance he will walk in 6 minutes?

A. 450 m C. 550 m

B. 360 m D. 480 m

14) The cost of using a car is $0.42 per minutes. Which of the following equations represents the total cost c, in dollars, for h hours of using the car?

A. $c = \frac{60h}{0.42}$ C. $c = 0.42\ (60h)$

B. $c = \frac{0.42}{60h}$ D. $c = 60h + 0.42$

15) Mary's average score after 5 tests is 72. What score on the 6th test would bring Mary's average up to exactly 75?

A. 77 C. 40

B. 90 D. 82

16) The equation $x^2 = 14x - 24$ has how many distinct real solutions?

A. 0 C. 2

B. 3 D. 1

Questions 17 to 19 refer to the following information.

A library has 500 books that include Mathematics, Physics, Chemistry, English, and History.

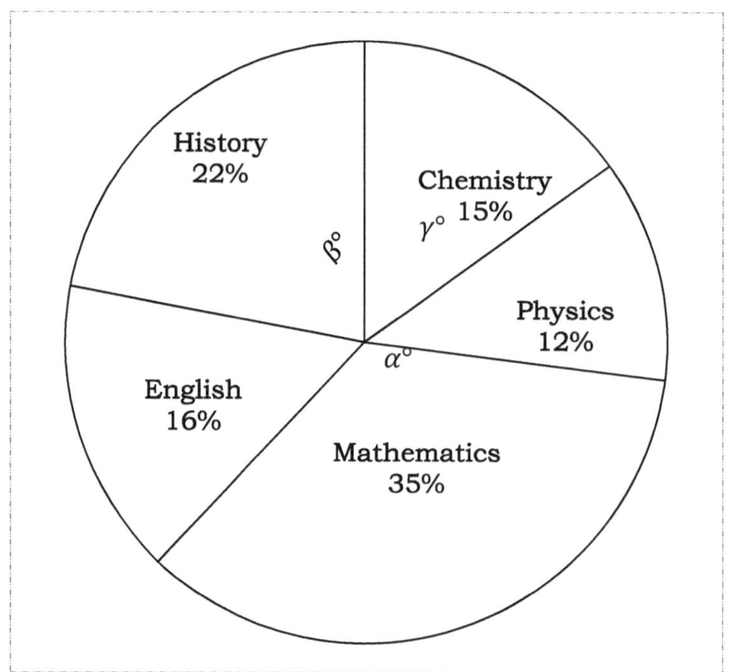

17) What is the product of the number of Mathematics and number of Chemistry books?

A. 23,125 C. 13,125

B. 18,351 D. 18,351

18) What are the values of angle α and β respectively?

A. 120°, 24° C. 54°, 79.2°

B. 72.5°, 65° D. 79°, 45°

19) The librarians decided to move some of the books in the Chemistry section to Mathematics section. How many books are in the Chemistry section if now $\alpha = \frac{3}{7}\gamma$?

A. 75 C. 95

B. 85 D. 25

20) In the xy – plane, line determined by the points $(3, m)$ and $(m, 6)$ passes through the origin. Which of the following could be the value of m?

A. $\sqrt{3}$

B. 6

C. $3\sqrt{2}$

D. 18

21) A function $g(1) = 3$ and $g(4) = 5$. A function $f(2) = 1$ and $f(5) = 16$ What is the value of $f(g(4))$?

A. 5 C. 1

B. 16 D. 3

22) What is the area of the following equilateral triangle if the side AB = 16 cm?

A. $64\sqrt{3}\ cm^2$ C. $16\sqrt{3}\ cm^2$

B. $32\sqrt{3}\ cm^2$ D. $16\ cm^2$

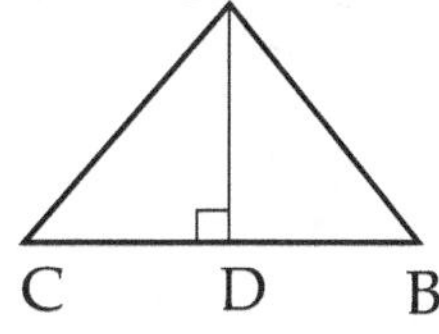

$$y = cx^2 + d\ , y = 6$$

23) In the system of equations above, c and d are constants. For which of the following values of c and d does the system of equations have exactly two real solutions?

A. $c = -4, d = 1$

B. $c = 2, d = 7$

C. $c = -1, d = 7$

D. $c = -2, d = -7$

24) From the figure below, which of the following must be true? (figure not drawn to scale)

A. $y = 3z$

B. $y = 4x$

C. $y \geq 2x$

D. $2y + 2x = z$

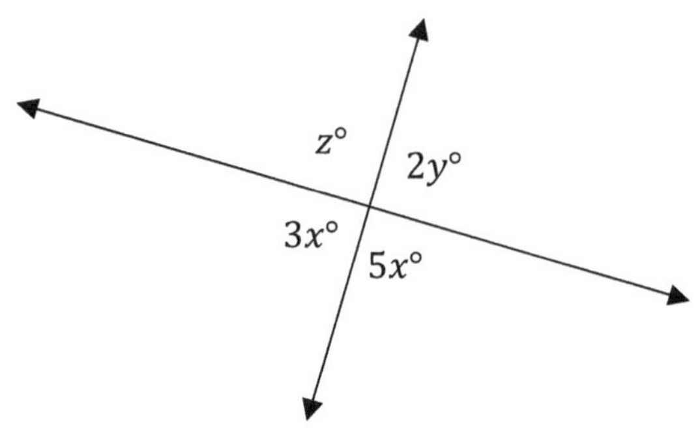

25) Point A lies on the line with equation $y - 6 = 2(x + 9)$. If the x –coordinate of A is 2, what is the y –coordinate of A?

A. 26

B. 30

C. 22

D. 28

26) A function $g(x)$ satisfies $g(1) = 3$ and $g(6) = 12$. A function $f(x)$ satisfies $f(6) = 18$ and $f(12) = 27$. What is the value of $f(g(6))$?

A. 18

B. 26

C. 27

D. 32

$$x^2 + y^2 + 2x + 6y = 2$$

27) The equation of a circle in the xy −plane is shown above. What is the radius of the circle?

A. 12

B. 10

C. $\sqrt{12}$

D. $\sqrt{10}$

$$\frac{3a - c}{a} = b$$

28) In the equation above, if c is positive and a is negative, which of the following must be true?

A. $b < 4$

B. $b > 3$

C. $b < -4$

D. $b > -3$

29) If $|a| < 3$, then which of the following is true? $(b > 0)$?

I. $-3b < ba < 3b$

II. $-a < a^2 < a$ if $a < 0$

III. $-8 < 2a - 2 < 4$

A. I only

B. II only

C. I and III only

D. I, II and III

30) Right triangle ABC is shown below. Which of the following is true for all possible values of angle A and B?

A. $\tan A = \tan B$

B. $\sin A = \cos B$

C. $\cot A = \cos B$

D. $\sin B = 1 - \cos A$

B c a A b C

Grid-ins Questions

Questions 31–38 are grid-ins questions. Solve the problems and enter your answers in the grid on the answer sheet as shown below.

31) If m is a positive integer and $\sqrt{4m+32} = m$, what is the value of m?

32) $f(a) = |16 - a^2|$, where x is a positive integer. If $f(a) = 20$, what is the value of a that satisfies the equation above?

33) In the standard (x, y) coordinate system plane, what is the area of the circle with the following equation? ($\pi = 3.14$) (Round the answer to one decimal place)

$$x^2 + y^2 + 6x - 8y + 7 = 0$$

34) The sum of four numbers is 900. One of the numbers, x is 25% more than the sum of the other three numbers. What is the value of x?

35) How many liters of a solution that is 40% salt must be added to 3 liters of a solution that is 60% salt to obtain a 25% salt solution?

36) The profit in dollars from a carwash is given by the function $P(x) = \frac{30a-480}{a} + b$, where a is the number of cars washed and b is a constant. If 30 cars were washed today for a total profit of \$450, what is the value of b?

$$f(x) = \frac{1}{(x+4)^2 + 6(x+4) + 9}$$

37) For what value of x is the function f(x) above undefinded?

38) In the following figure, ABCD is a rectangle, and E and F are points on AD and DC, respectively and DE = 10 and DF = 5. The area of ΔBED is 45, and the area of ΔBDF is 40. What is the perimeter of the rectangle?

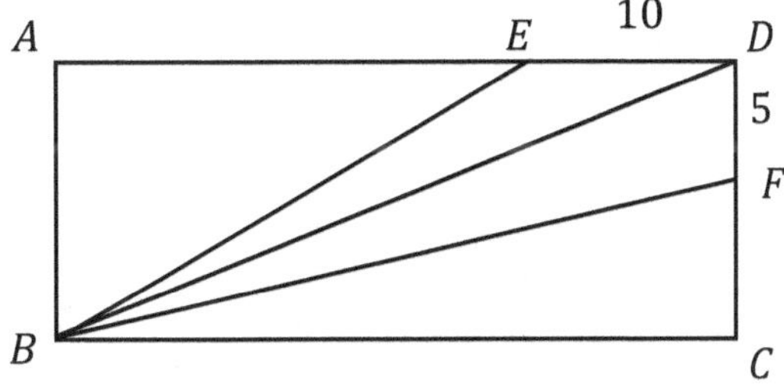

STOP

This is the End of this Section. You may check your work on this section if you still have time.

SAT Math Practice Test 2

Section 1

- ❖ **20 Questions.**
- ❖ **Total time for this test: 25 Minutes.**
- ❖ **You may NOT use a calculator on this Section.**

Administered *Month Year*

1) For $i = \sqrt{-1}$, which of the following is equivalent of $\frac{-2+5i}{1+2i}$?

A. $\frac{-8+9i}{15}$

C. $\frac{8+9i}{5}$

B. $\frac{4-9i}{25}$

D. $\frac{8-9i}{25}$

2) If function is defined as $f(x) = bx^2 + 5x$, and b is a constant and $f(5) = 75$.

What is the value of $f(4)$?

A. 28

C. 52

B. 57

D. 54

3) Find the solution (x, y) to the following system of equations?

$$2x + 4y = -8$$

$$8x + 15y = -30$$

A. $(-2, 2)$

C. $(2, -2)$

B. $(-2, 0)$

D. $(0, -2)$

4) Calculate $f(5)$ for the function $f(x) = x^2 - 4x + 8$.

A. 45

C. 8

B. 53

D. 13

5) If $\frac{x-7}{5} = N$ and $N = 6$, what is the value of x?

A. 49

C. 23

B. 42

D. 37

6) Which of the following is equal to $b^{\frac{5}{7}}$?

A. $\sqrt{b^{\frac{7}{5}}}$

B. $b^{\frac{7}{5}}$

C. $\sqrt[7]{b^5}$

D. $\sqrt[5]{b^7}$

7) On Saturday, Sara read N pages of a book each hour for 6 hours, and Mary read M pages of a book each hour for 11 hours. Which of the following represents the total number of pages of book read by Sara and Mary on Saturday?

A. $66MN$

B. $6N + 11M$

C. $1.83MN$

D. $11N + 6M$

8) What is the sum of all values of n that satisfies $2n^2 + 26n + 80 = 0$?

A. 40

B. -3

C. -13

D. -16

$$y = x^2 - 8x + 15$$

9) The equation above represents a parabola in the xy-plane. Which of the following equivalent forms of the equation displays the x-intercepts of the parabola as constants or coefficients?

A. $y = x + 3$

B. $y = 3x(x - 5)$

C. $y = (x + 3)(x + 5)$

D. $y = (x - 3)(x - 5)$

10) What is the value of $\frac{21b}{c}$ when $\frac{c}{b} = 7$

A. 3

B. 7

C. 14

D. 0.3

11) Which of the following is equivalent to $\frac{3x+(4x)^2+(5x)^3}{x}$?

A. $25x^2 + 4x + 3$

B. $125x^2 + 16x + 3$

C. $125x^2 + 16x$

D. $125x^3 + 16x^2 + 3$

12) If $\frac{a-b}{b} = \frac{5}{12}$, then which of the following must be true?

A. $\frac{a}{b} = \frac{12}{5}$

B. $\frac{a}{b} = \frac{17}{12}$

C. $\frac{a}{b} = \frac{7}{12}$

D. $\frac{a}{b} = \frac{13}{12}$

13) Which of the following lines is parallel to $16y - 4x = 48$?

A. $y = \frac{1}{4}x + 12$

B. $y = 4x + 1$

C. $y = -4x - 1$

D. $y = 4x - 8$

14) The function $g(x)$ is defined by a polynomial. Some values of x and $g(x)$ are shown in the table below. Which of the following must be a factor of $g(x)$?

x	$g(x)$
0	−6
3	−3
6	0

A. $4x - 7$

B. $x - 9$

C. $x - 6$

D. $5x - 9$

15) If $x■y = \sqrt{x^2 + 4y}$, what is the value of $4■5$?

A. $\sqrt{108}$

B. 6

C. 4

D. 5

Grid-ins Questions

Questions 16–20 are grid-ins questions. Solve the problems and enter your answers in the grid on the answer sheet as shown below.

16) Angle a is 450 degrees and can be written $x\pi$ in radian. What is the value of x?

17) A construction company is building a wall. The company can build 40 cm of the wall per minute. After 55 minutes $\frac{4}{5}$ of the wall is completed. How many meters is the wall?

18) If the ratio of $5a$ to $9b$ is $\frac{1}{45}$, what is the ratio of a to b?

19) If the interior angles of a quadrilateral are in the ratio 2:3:6:9, what is the measure of the largest angle?

20) In a right triangle ABC, angle B is a right angle. If $\tan \mathrm{A} = \frac{9}{12}$, what is $\cos C$?

STOP

This is the End of this Section. You may check your work on this section if you still have time.

SAT Math Practice Test 2

Section 2

- **38 Questions.**
- **Total time for this test: 55 Minutes.**
- **You may NOT use a calculator on this Section.**

Administered Month Year

1) If $\alpha = 3\beta$ and $\beta = 5\gamma$, how many α are equal to 45γ?

A. 15

B. 5

C. 3

D. 4

2) If $a - b > 2$ and $a + b < 12$, which of the following pairs could not be the values of a and b?

A. (7, 3)

B. (8, 5)

C. (4, 0)

D. (6, 2)

3) What is the value of x in the following system of equations?

$$x + 3y = 14$$

$$5x + 8y = 49$$

A. $x = 5$

B. $x = -5$

C. $x = 3$

D. $x = -3$

4) In a hotel, there are 17 floors and x rooms on each floor. If each room has exactly y chairs, which of the following gives the total number of chairs in the hotel?

A. $17xy$

B. $34xy$

C. $17x + y$

D. $x + 17y$

5) When 7 times the number x is added to 4, the result is 53. What is the result when 3 times x is added to 8?

A. 19

B. 29

C. 26

D. 31

6) If $9h + g = 11h + 5$, what is g in terms of h?

A. $h = 2g - 5$

C. $h = 20g$

B. $g = 2h + 5$

D. $g = h + 5$

7) If $4 + 5x$ is 12 more than 27, what is the value of $15x$?

A. 117

C. 95

B. 70

D. 105

x	2	3	4
$g(x)$	-9	-14	-19

8) The table above shows some values of linear function $g(x)$. Which of the following defines $g(x)$?

A. $g(x) = x + 5$

C. $g(x) = -5x + 1$

B. $g(x) = x - 11$

D. $g(x) = -5x - 1$

9) In a certain set of numbers, the ratio of integers to non-integers is 9:11. What percent of the numbers in the set are integers?

A. 90%

C. 5%

B. 10%

D. 50%

10) What is the y –intercept of the line with the equation $x - 3y = 18$?

A. 6

C. 18

B. -6

D. -18

$$8x^2 + 5x - 11 \ , \ 3x^2 - 4x + 15$$

11) Which of the following is the sum of the two polynomials shown above?

A. $4x^2 + 11x + 1$

B. $4x^2 + 11x + 11$

C. $11x^2 + x + 4$

D. $11x^2 + 4x + 1$

Questions 12 to 14 are based on the following data.

The result of a research shows the number of men and women in four cities of a country.

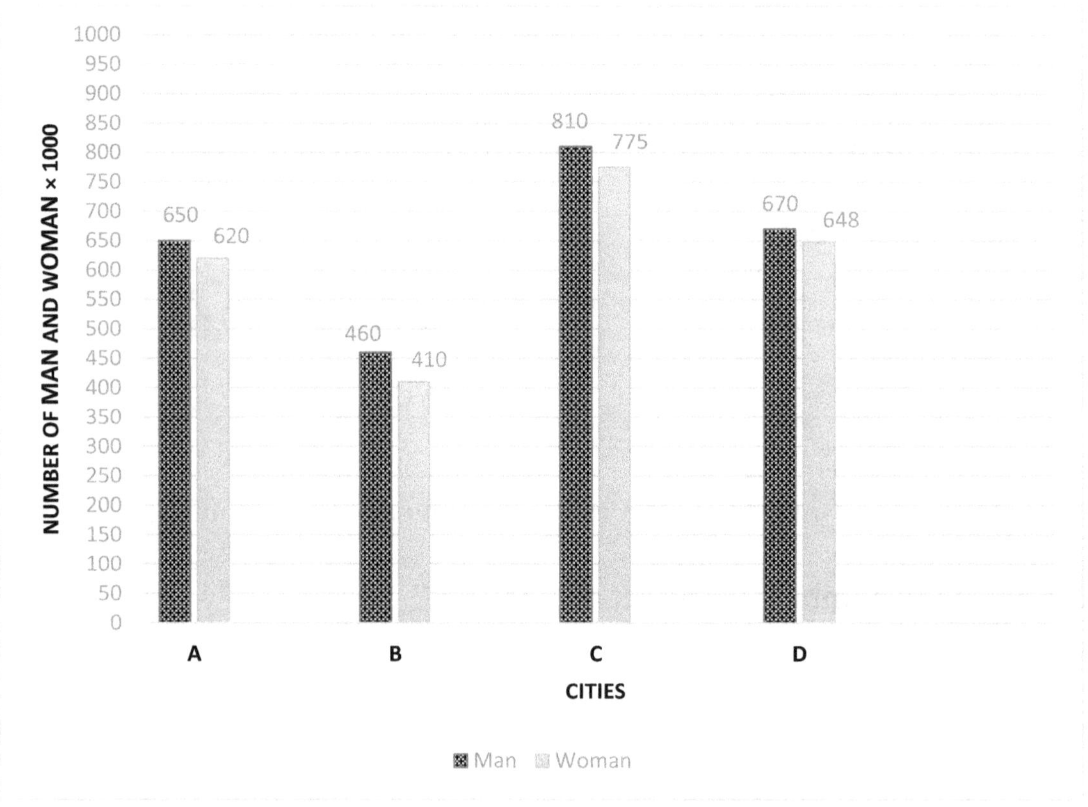

12) What's the ratio of percentage of men in city C to percentage of men in city B?

A. 0.65

B. 0.07

C. 0.97

D. 96.65

13) What's the minimum ratio of woman to man in the four cities?

A. 0.83

C. 0.89

B. 0.77

D. 0.86

14) How many women should be added to city D until the ratio of women to men will be 1.7?

A. 480

C. 431

B. 491

D. 498

15) If $8a - 7 = 33$ what is the value of $5a$?

A. 5

C. 25

B. 20

D. 30

16) If $x \neq 0$ and $x = x^{-3}$, what is the value of x?

A. -3

C. -1

B. 1

D. 3

17) Which of the following is equal to expression $\frac{8}{x^4} + \frac{3x-4}{x^5}$?

A. $\frac{11x-4}{x^5}$

C. $\frac{-3x-8}{x^5}$

B. $\frac{8x+3}{x^5}$

D. $\frac{8x^2+3x-4}{x^6}$

18) What is the average of $-3x + 6, -5x - 8$ and $5x + 14$?

A. $x + 8$

C. $4x + 8$

B. $2x - 8$

D. $-x + 4$

19) The system of equations below has solution (x, y). What is the value of x?

$$\frac{5}{2}y = 6$$

$$x - \frac{5}{2}y = 2$$

A. 12

B. 6

C. 8

D. 2

20) Which of the following is the equation of a quadratic graph with a vertex $(4, -4)$?

A. $y = 4x^2 - 4$

B. $y = -4x^2 + 4$

C. $y = x^2 + 6x - 8$

D. $y = 5(x - 4)^2 - 4$

21) A rectangle was altered by increasing its length by 22 percent and decreasing its width by s percent. If these alterations decreased the area of the rectangle by 8.5 percent, what is the value of s.

A. 20

B. 45

C. 25

D. 35

22) If $b^{\frac{-7}{2}} = x$, where $b > 0, x > 0$. Which of the following equations gives b in term of x?

A. $7\sqrt{x}$

B. $\sqrt[7]{\frac{1}{x^2}}$

C. $\frac{1}{x^2}$

D. $\sqrt{x^7}$

23) Tickets for a talent show cost \$2 for children and \$5 for adults. If John spends at least \$23 but no more than \$27 on x children's tickets and 3 adult ticket, what are two possible values of x?

A. 4, 2
C. 1, 3
B. 3, 4
D. 4, 1

24) Sara orders a box of pen for \$3 per box. A tax of 3.8% is added to the cost of the pens before a flat shipping fee of \$8 closest out the transaction. Which of the following represents total cost of p boxes of pens in dollars?

A. $1.038(3p) + 8$
C. $1.038(8p) + 3$
B. $8p + 3$
D. $3p + 38$

25) A plant grows at a linear rate. After nine weeks, the plant is 63 cm tall. Which of the following functions represents the relationship between the height (y) of the plant and number of weeks of growth (x)?

A. $y(x) = 63x + 7$
C. $y(x) = 63x$
B. $y(x) = 7x + 63$
D. $y(x) = 7x$

26) A central angle of a circle measures 120 degrees. If its corresponding arc measures 6 unit, what is the area of the circle?

A. $A = 9\pi$
C. $A = \frac{81}{\pi}$
B. $A = 81\pi$
D. $A = \frac{81}{\pi^2}$

27) In four successive hours, a car travels 52 km, 44 km, 32 km, and 59 km. In the next six hours, it travels with an average speed of 51 km per hour. Find the total distance the car traveled in 10 hours.

A. 463 km

B. 480 km

C. 493 km

A. 590 km

28) The table below shows some values for the function $g(x)$. If $g(x)$ is a linear function, what is the value of a in terms of b?

A. $a = -3b - 7$

B. $b = -5a + 14$

C. $b = -5a - 4$

D. $a = -3b - 9$

x	$g(x)$
a	b
1	−9
−4	16

29) In the xy −plane, point (2, 4) lies on the graph of the function $g(x) = 2x^2 + cx - 8$. What is the value of c?

A. 2

B. −2

C. 4

D. 3

30) For what value of x is $|x - 9| + 9$ equal to 0?

A. 9

B. no value of x

C. −9

D. 18

Grid-ins Questions

Questions 31–38 are grid-ins questions. Solve the problems and enter your answers in the grid on the answer sheet as shown below.

31) One of the zeros of the function $f(x) = 3x^3 - 39x^2 + 120x$ is zero. What is one of the other zeros of this function?

32) A restaurant sells salads for $10 each and drinks for $5 each. The restaurant's revenue from selling a total of 400 salads and drinks in two days was $2,100. How many drinks were sold in total?

33) The mean of 40 test scores was calculated as 65. But it turned out that one of the scores was misread as 75 but it was 35. What is the mean?

34) A certain experiment has 4 possible mutually exclusive outcomes and have probabilities $n, \frac{n}{4}, \frac{n}{8}, \frac{n}{16}$, respectively. What is the value of n?

35) In the following sequence 4 is the first term. What is the value of the 31st term of the sequence?

$$4, 10, 16, 22, \ldots$$

36) Two cars are 130 miles apart. They both drive in a straight line toward each other. If Car A drives at 50 mph and Car B drives at 80 mph, then how many miles apart will they be exactly 30 minutes before they meet?

37) If $f(x) = \frac{2x-7}{3}$ and $f^{-1}(x)$, is the inverse of $f(x)$, what is the value of $f^{-1}(5)$?

38) Two acute angles are shown below and $\cos(a) = \sin(b)$. If $a = 2n + 3$ and $b = 5n - 11$, what is the value of n?

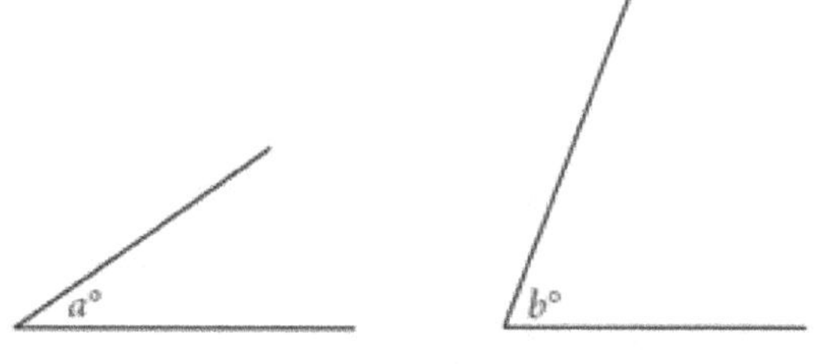

STOP

This is the End of this Section. You may check your work on this section if you still have time.

Chapter 17 :

Answers and Explanations

Answer Key

※ Now, it's time to review your results to see where you went wrong and what areas you need to improve!

SAT Math Practice Test 1

Section 1			
1	A	16	12
2	C	17	1.12
3	C	18	5
4	C	19	6
5	C	20	248
6	C		
7	A		
8	B		
9	B		
10	D		
11	B		
12	C		
13	A		
14	D		
15	B		

Section 2					
1	D	16	C	31	8
2	A	17	C	32	6
3	C	18	C	33	56.52
4	C	19	A	34	500
5	C	20	C	35	7
6	C	21	B	36	436
7	D	22	A	37	−7
8	D	23	C	38	50
9	D	24	D		
10	D	25	D		
11	B	26	C		
12	C	27	C		
13	A	28	B		
14	C	29	C		
15	B	30	B		

Answers and Explanations

SAT Math Practice Test 2

Section 1			
1	C	16	2.5
2	C	17	27.5
3	D	18	0.04
4	D	19	162
5	D	20	$\frac{4}{5}$
6	C		
7	B		
8	C		
9	D		
10	A		
11	B		
12	B		
13	A		
14	C		
15	B		

Section 2					
1	C	16	B	31	$5\ or\ 8$
2	B	17	A	32	380
3	A	18	D	33	64
4	A	19	C	34	$\frac{16}{23}$
5	B	20	D	35	184
6	B	21	C	36	65
7	D	22	B	37	11
8	C	23	B	38	14
9	C	24	A		
10	B	25	D		
11	C	26	C		
12	D	27	C		
13	C	28	C		
14	B	29	A		
15	C	30	B		

SAT Mathematics

Practice Tests 1:

Section 1

1) Answer: A.

If $f(x) = 4x + 3(2x + 5) - 7$, then find $f(2x)$ by substituting $2x$ for every x in the function. This gives: $f(2x) = 4(2x) + 3(2(2x) + 5) - 7$,

It simplifies to: $f(2x) = 4(2x) + 3(2(2x) + 5) - 7 = 8x + 12x + 15 - 7 = 20x + 8$

2) Answer: C.

First, find the equation of the line. All lines through the origin are of the form $y = mx$, so the equation is $y = \frac{1}{6}x$. Of the given choices, only choice C $(18, 3)$, satisfies this equation: $y = \frac{1}{6}x \rightarrow 3 = \frac{1}{6}(18) = 3$

3) Answer: C.

Method 1: Plugin the values of x and y provided in the options into both equations.

A. $(-1, 3)$ $\quad x + 2y = 0 \rightarrow -1 + 2(3) = 5 \neq 0$

B. $(3, -4)$ $\quad x + 2y = 0 \rightarrow 3 - 2(-4) = 11 \neq 0$

C. $(2, -1)$ $\quad x + 2y = 0 \rightarrow 2 + 2(-1) = 0 = 0$

D. $(-6, -4)$ $\quad x + 2y = 0 \rightarrow -6 + 2(-4) = -14 \neq 0$

Only option C is correct.

Method 2: Multiplying each side of $x + y = 0$ by 4 gives $3x + 3y = 0$. Then, adding the corresponding side of $3x + 3y = 0$ and $5x - 3y = 16$ gives $8x = 16$. Dividing each side of $8x = 16$ by 8 gives $x = 2$. Finally, substituting 2 for x in $x + 2y = 0$, or $y = -1$. Therefore, the solution to the given system of equations is $(2, -1)$.

4) Answer: D.

Add 12 both sides of the equation $6x - 12 = 24$ gives $6x = 24 + 12 = 36$.

Dividing each side of the equation $6x = 36$ by 6 gives $x = 6$. Substituting 6 for x in the expression $8x - 20$ gives $8(6) - 20 = 28$.

5) Answer: C.

$(12n^2 + 4n + 6) - (7n^2 - 2)$

Add like terms together: $12n^2 - 7n^2 = 5n^2$

$4n$ doesn't have like terms.$6 - (-2) = 8$

Combine these terms into one expression to find the answer: $5n^2 + 4n + 8$

6) Answer: C.

You can find the possible values of a and b in $(ax + 7)(bx + 5)$ by using the given equation $a + b = 12$ and finding another equation that relates the variables a and b. Since $(ax + 7)(bx + 5) = 27x^2 + cx + 35$, expand the left side of the equation to obtain $abx^2 + 7bx + 5ax + 35 = 27x^2 + cx + 35$

Since ab is the coefficient of x^2 on the left side of the equation and 27 is the coefficient of x^2 on the right side of the equation, it must be true that $ab = 27$

The coefficient of x on the left side is $7b + 5a$ and the coefficient of x in the right side is c. Then: $7b + 5a = c$; $a + b = 12$, then: $a = 12 - b$

Now, plug in the value of a in the equation $ab = 27$. Then:

$ab = 27 \rightarrow (12 - b)b = 27 \rightarrow 12b - b^2 = 27$

Add $-12b + b^2$ both sides. Then: $b^2 - 12b + 27 = 0$

Solve for b using the factoring method. $b^2 - 12b + 27 = 0 \rightarrow (b - 3)(b - 9) = 0$

Thus, either $b = 3$ and $a = 9$, or $b = 9$ and $a = 3$. If $b = 3$ and $a = 9$, then, $7b + 5a = c \rightarrow 7(3) + 5(9) = c \rightarrow c = 66$

If $b = 9$ and $a = 3$, then $7b + 5a = c \rightarrow 7(9) + 5(3) = c \rightarrow c = 78$

Therefore, the two possible values for c are 66 and 78.

7) Answer: A.

To rewrite $\frac{1}{\frac{1}{x-5}+\frac{1}{x+6}}$, first simplify $\frac{1}{x-5} + \frac{1}{x+6}$.

$\frac{1}{x-5} + \frac{1}{x+6} = \frac{1(x+6)}{(x-5)(x+6)} + \frac{1(x-5)}{(x+6)(x-5)} = \frac{(x+6)+(x-5)}{(x+6)(x-5)}$;

Then: $\frac{1}{\frac{1}{x-5}+\frac{1}{x+6}} = \frac{1}{\frac{(x+6)+(x-5)}{(x+6)(x-5)}} = \frac{(x-5)(x+6)}{(x-5)+(x+6)}$. (Remember, $\frac{1}{\frac{1}{x}} = x$)

This result is equivalent to the expression in choice A.

8) Answer: B.

Since (0, 0) is a solution to the system of inequalities, substituting 0 for x and 0 for y in the given system must result in two true inequalities. After this substitution, $y > b + x$ becomes $0 > b$, and $y < a + x$ becomes $0 < a$. Hence, b is negative and a is positive. Therefore, $b < a$.

9) Answer: B.

To solve this problem, first recall the equation of a line: $y = mx + b$

Where $m = slope$

$y = y - intercept$

Remember that slope is the rate of change that occurs in a function and that the y −intercept is the y value corresponding to $x = 0$.

Since the height of John's plant is 3 inches tall when he gets it. Time (or x) is zero. The plant grows 8 inches per year. Therefore, the rate of change of the plant's height is 8. The y −intercept represents the starting height of the plant which is 3 inches.

10) Answer: D.

First find the slope of the line using the slope formula: $m = \frac{y_2 - y_1}{x_2 - x_1}$

Substituting in the known information. $(x_1, y_1) = (2, -2)$; $(x_2, y_2) = (4, -8)$

$m = \frac{-8-(-2)}{4-2} = \frac{-6}{2} = -3$

Now the slope to find the equation of the line passing through these points: $y = mx + b$

Choose one of the points and plug in the values of x and y in the equation to solve for b. Let's choose point $(2, -2)$. Then:

$y = mx + b \rightarrow -2 = -3(2) + b \rightarrow -2 = -6 + b \rightarrow b = -2 + 6 = 4$

The equation of the line is: $y = -3x + 4$

Now, plug in the points provided in the choices into the equation of the line.

A. $(-1, 5)$; $y = -3x + 4 \rightarrow 5 = -3(-1) + 4 \rightarrow 7 = 1$ This is NOT true.

B. $(1, 3)$; $y = -3x + 4 \rightarrow 3 = -3(1) + 4 \rightarrow 3 = 1$ This is NOT true.

C. $(2, 6)$; $y = -3x + 4 \rightarrow 6 = -3(2) + 4 \rightarrow 6 = -2$ This is NOT true.

D. $(0, 4)$; $y = -3x + 4 \rightarrow 4 = -3(0) + 4 \rightarrow 4 = 4$ This is true!

Therefore, the only point from the choices that lies on the line is $(0, 4)$.

11) Answer: B.

The input value is 3. Then: $x = 3$

$f(x) = 3x^3 - 4x + 5 \rightarrow f(3) = 3(3)^3 - 4(3) + 5 = 81 - 12 + 5 = 74$

12) Answer: C.

Multiplying each side of $\frac{3}{x} = \frac{18}{x-15}$ by $x(x - 15)$ gives $3(x - 15) = 18(x)$, distributing the 3 over the values within the parentheses yields $x - 15 = 6x$ or $x = -3$.

Therefore, the value of $\frac{x}{3} = \frac{-3}{3} = -1$.

13) Answer: A.

In order to figure out what the equation of the graph is, fist find the vertex. From the graph we can determine that the vertex is at $(-3, 1)$.

We can use vertex form to solve for the equation of this graph.

Recall vertex form, $y = a(x - h)^2 + k$, where h is the x coordinate of the vertex, and k is the y coordinate of the vertex. Plugging in our values, you get $y = a(x + 3)^2 + 1$

To solve for a, we need to pick a point on the graph and plug it into the equation.

Let's pick $(0, 10)$, $10 = a(0 + 3)^2 + 1 \Rightarrow 10 = a(3)^2 + 1 \Rightarrow 10 = 9a + 1 \Rightarrow 9a = 9 \Rightarrow a = 1$;

Now the equation is: $y = 1(x + 3)^2 + 1$; Let's expand this, $y = (x^2 + 6x + 9) + 1$, $y = x^2 + 6x + 10$. The equation in Choice D is the same.

14) Answer: D.

The equation of a circle can be written as $(x - h)^2 + (y - k)^2 = r^2$

where (h, k) are the coordinates of the center of the circle and r is the radius of the circle. Since the coordinates of the center of the circle are (3, 5), the equation is $(x - 3)^2 + (y - 5)^2 = r^2$, where r is the radius. The radius of the circle is the distance from the center (3, 5), to the given endpoint of a radius, $\left(\frac{5}{2}, 8\right)$. By the distance formula, $r^2 = \left(\frac{5}{2} - 3\right)^2 + (8 - 5)^2 = \frac{1}{4} + 9 = \frac{37}{4}$

Therefore, an equation of the given circle is $(x - 3)^2 + (y - 5)^2 = \frac{37}{4}$

15) Answer: B.

To solve for $\cos A$ first identify what is known.

The question states that ΔABC is a right triangle whose $\angle B = 90^{\circ}$ and $\sin C = \frac{3}{5}$.

It is important to recall that any triangle has a sum of interior angles that equals 180 degrees. Therefore, to calculate $\cos A$ use the complimentary angles identify of trigonometric function.

$\cos A = \cos(90 - C)$, Then: $\cos A = \sin C$

For complementary angles, sin of one angle is equal to cos of the other angle. $\cos A = \frac{3}{5}$

16) Answer: 12.

$x^2 + 8x + r = (x + 6)(x + p) = x^2 + (6 + p)x + 6p$

On the left side of the equation the coefficient of x is 8 and on the right side of the equation the coefficient of x is $6 + p$. Thus $6 + p = 8 \rightarrow p = 2$ *and* $r = 6p = 6(2) = 12$

17) Answer: 1.12.

The initial deposit earns 12 percent interest compounded annually. Thus, at the end of year 1, the new value of the account is the initial deposit of \$140 plus 12 percent of the initial deposit: $\$140 + \frac{12}{100}(\$140) = \$140(1.12)$.

Since the interest is compounded annually, the value at the end of each succeeding year is the sum of the previous year's value plus 12 percent of the previous year's value. This is equivalent to multiplying the previous year's value by 1.12. Thus, after 2 years, the value will be $\$140(1.12)\,(1.12) = \$(140)(1.12)^2$; and after 3 years, the value will be $(140)(1.12)^3$; and after n years, the value will be $(140)(1.12)^n$. Therefore, in the formula for the value for Sara's account after n years $(140)(x)^n$, the value of x is 1.12.

18) Answer:5.

One liter = 1,000 $cm^3 \rightarrow$ 7 liters = 7,000 cm^3

$7{,}000 = 70 \times 20 \times h \rightarrow h = \frac{7{,}000}{1{,}400} = 5$ cm

19)Answer: 6.

The four-term polynomial expression can be factored completely, by grouping, as follows:

$(x^3 - 6x^2) + (3x - 18) = 0$

$x^2(x - 6) + 3(x - 6) = 0$

$(x - 6)(x^2 + 3) = 0$

By the zero-product property, set each factor of the polynomial equal to 0 and solve each resulting equation for x. This gives $x = 6 \; or \; x = \pm i\sqrt{3}$, respectively. Because the equation the question asks for the real value of x that satisfies the equation, the correct answer is 6.

20)Answer: 248.

Let L be the length of the rectangular and W be the width of the rectangular. Then, $L = 3W + 7$

The perimeter of the rectangle is 78 meters. Therefore: $2L + 2W = 78, L + W = 39$

Replace the value of L from the first equation into the second equation and solve for W: $(3W + 7) + W = 39 \rightarrow 4W + 7 = 39 \rightarrow 4W = 32 \rightarrow W = 8$

The width of the rectangle is 8 meters, and its length is: $L = 3W + 7 = 3(8) + 7 = 31$

The area of the rectangle is: length × width = 31 × 8 = 248

SAT Mathematics

Practice Tests 1:

Section 2

1) Answer: D.

Substituting 4 for x and 14 for y in $y = nx + 6$ gives $14 = (n)(4) + 6$, which gives $n = 2$. Hence, $y = 2x + 6$. Therefore, when $x = 5$, the value of y is: $y = (2)(5) + 6 = 16$.

2) Answer: A.

Subtracting $4x$ and adding 3 to both sides of $4x - 9 \geq 5x - 3$ gives $-6 \geq x$. Therefore, x is a solution to $4x - 9 \geq 5x - 3$ if and only if x is less than or equal to -6 and x is NOT a solution to $4x - 9 \geq 5x - 3$ if and only if x is greater than -6. Of the choices given, only -5 is greater than -6 and, therefore, cannot be a value of x.

3) Answer: C.

Given the two equations, substitute the numerical value of a into the second equation to solve for x. $a = \sqrt{2}$, $6a = \sqrt{6x}$

Substituting the numerical value for a into the equation with x is as follows.

$6(\sqrt{2}) = \sqrt{6x}$, From here, distribute the 6. $6\sqrt{2} = \sqrt{6x}$

Now square both side of the equation. $(6\sqrt{2})^2 = (\sqrt{6x})^2$

Remember to square both terms within the parentheses. Also, recall that squaring a square root sign cancels them out.

$6^2\sqrt{2}^2 = 6x$, $36(2) = 6x$, $72 = 6x$, $x = 12$

4) Answer: C.

First square both sides of the equation to get $9m - 20 = m^2$

Subtracting both sides by $9m - 20$ gives us the equation $m^2 - 9m + 20 = 0$

Here you can solve the quadratic equation by factoring to get $(m - 4)(m - 5) = 0$

For the expression $(m - 4)(m - 5)$ to equal zero, $m = 4$ or $m = 5$

5) Answer: C.

Adding 18 to each side of the inequality $4n - 13 \geq 2$ yields the inequality $4n + 5 \geq 20$. Therefore, the least possible value of $4n + 5$ is 20.

6) Answer: C.

Since $f(x)$ is linear function with a negative slop, then when $x = -1, f(x)$ is maximum and when $x = 3, f(x)$ is minimum. Then the ratio of the minimum value to the maximum value of the function is: $\frac{f(3)}{f(-1)} = \frac{-5(3)+3}{-5(-1)+2} = \frac{-12}{7} = -1\frac{5}{7}$

7) Answer: D.

There can be 0, 1, or 2 solutions to a quadratic equation. In standard form, a quadratic equation is written as: $ax^2 + bx + c = 0$

For the quadratic equation, the expression $b^2 - 4ac$ is called discriminant. If discriminant is positive, there are 2 distinct solutions for the quadratic equation. If discriminant is 0, there is one solution for the quadratic equation and if it is negative the equation does not have any solutions.

To find number of solutions for $x^2 = 6x - 9$, first, rewrite it as $x^2 - 6x + 9 = 0$.

Find the value of the discriminant. $b^2 - 4ac = (-6)^2 - 4(1)(9) = 36 - 36 = 0$

Since the discriminant is zero, the quadratic equation has one distinct solution.

8) Answer: D.

Of the 40 employees, there are 16 females under age 45 and 14 males age 45 or older. Therefore, the probability that the person selected will be either a female under age 45 or a male age 45 or older is: $\frac{16}{50} + \frac{14}{50} = \frac{30}{50} = \frac{3}{5}$

9) Answer: D.

To solve the equation for y, multiply both sides of the equation by the reciprocal of $\frac{8}{3}$, which is $\frac{3}{8}$, this gives $\left(\frac{3}{8}\right) \times \frac{8}{3}y = \frac{16}{5} \times \left(\frac{3}{8}\right)$, which simplifies to $y = \frac{48}{40} = \frac{6}{5}$.

10) Answer: D.

In the figure angle A labeled $(2x + 10)$ and it measures 30. Thus, $2x + 10 = 30$ and $2x = 20$ or $x = 10$.

That means that angle B, which is labeled $(8x)$, must measure $8 \times 10 = 80$.

Since the three angles of a triangle must add up to 180, $30 + 80 + y - 12 = 180$, then:

$y + 110 - 12 = 180 \rightarrow y = 180 - 98 = 82$

11) Answer: B.

A zero of a function corresponds to an x-intercept of the graph of the function in the xy-plane. Therefore, the graph of the function $g(x)$, which has three distinct zeros, must have three x –intercepts. Only the graph in choice B has three x –intercepts.

12) Answer: C.

The fastest way to find the answer is to pick numbers. Pick a number for x that has a remainder of 5 when divided by 6, such as 35. Increase the number you picked by 8. In this case $35 + 8 = 43$. Now divide 43 by 6, which gives you remainder 1. Therefore, the answer is 1.

13) Answer: A.

Because Jack walks 50 meters in 40 seconds, and 6 minutes is equal to 360 seconds, use the proportion to solve. $\frac{50\ meters}{40\ sec} = \frac{x\ meters}{360\ sec}$

The proportion can be simplified to $\frac{50}{40} = \frac{x}{360}$ then each side of the equation can be multiplied by 360, giving $\frac{(360)(50)}{40} = x = 450$.

Therefore, 450 meters is the distance Jack will walk in 6 minutes.

14) Answer: C.

\$0.42 per minute to use car. This per-minute rate can be converted to the hourly rate using the conversion 1 hour = 60 minutes, as shown below.

$\frac{0.42}{minute} \times \frac{60\ minutes}{1\ hours} = \frac{\$(0.38 \times 60)}{hour}$; Thus, the car costs $\$(0.42 \times 60)$ per hour.

Therefore, the cost c, in dollars, for h hours of use is $c = (0.42 \times 60)h$,

Which is equivalent to $c = 0.42(60h)$

15) Answer: B.

The best way to deal with changing averages is to use the sum.

Use the old average to figure out the total of the first 5 scores:

Sum of first 5 scores: $(5)(72) = 360$

Use the new average to figure out the total she needs after the 5th score:

Sum of 6 score: $(6)(75) = 450$

To get her sum from 360 to 450, Mary needs to score $450 - 360 = 90$.

16) Answer: C.

To solve a quadratic equation, put it in the $ax^2 + bx + c = 0$ form, factor the left side, and set each factor equal to 0 separately to get the two solutions.

To solve $x^2 = 14x - 24$, first, rewrite it as $x^2 - 14x + 24 = 0$.

Then factor the left side: $x^2 - 14x + 24 = 0$, $(x - 12)(x - 2) = 0$

$x = 2$ or $x = 12$, There are two solutions for the equation.

17) Answer: C.

Number of Mathematics book: $0.15 \times 500 = 75$

Number of Chemistry book: $0.35 \times 500 = 175$

Product of number of Mathematics and number of Chemistry books: $75 \times 175 = 13{,}125$

18) Answer: C.

The angle γ is: $0.15 \times 360 = 54°$

The angle β is: $0.22 \times 360 = 79.2°$

19) Answer: A.

According to the graph, 50% of the books are in the Mathematics and Chemistry sections. Therefore, there are 250 books in these two sections: $0.50 \times 500 = 250$

$\gamma + \alpha = 250$, and $\gamma = \frac{3}{7}\alpha$ (Replace γ by $\frac{3}{7}\alpha$ in the first equation)

$\gamma + \alpha = 250 \rightarrow \frac{3}{7}\alpha + \alpha = 250 \rightarrow \frac{10}{7}\alpha = 250$

$\rightarrow$multiply both sides by $\frac{7}{10}$: $\left(\frac{7}{10}\right)\frac{10}{7}\alpha = 250 \times \left(\frac{7}{10}\right) \rightarrow \alpha = \frac{250\times7}{10} = 175$

$\alpha = 175 \rightarrow \gamma = \frac{3}{7}\alpha \rightarrow \gamma = \frac{3}{7} \times 175 = 75$

There are 75 books in the Chemistry section.

20) Answer: C.

The line passes through the origin, $(3, m)$ and $(m, 6)$.

Any two of these points can be used to find the slope of the line. Since the line passes through (0, 0) and $(3, m)$, the slope of the line is equal to $\frac{m-0}{3-0} = \frac{m}{3}$. Similarly, since the line passes through (0, 0) and $(m, 6)$, the slope of the line is equal to $\frac{6-0}{m-0} = \frac{6}{m}$. Since each expression gives the slope of the same line, it must be true that $\frac{m}{3} = \frac{6}{m}$

Using cross multiplication gives

$\frac{m}{3} = \frac{6}{m} \rightarrow m^2 = 18 \rightarrow m = \pm\sqrt{18} = \pm\sqrt{9 \times 2} = \pm\sqrt{9} \times \sqrt{2} = \pm 3\sqrt{2}$

21) Answer: B.

It is given that $g(4) = 5$. Therefore, to find the value of $f(g(4))$, then $f(g(4)) = f(5) = 16$

22) Answer: A.

Area of the triangle is: $\frac{1}{2}\ AD \times DC$ and AD is perpendicular to BC.

Triangle ADC is a $30° - 60° - 90°$ right triangle. The relationship among all sides of right triangle $30° - 60° - 90°$ is provided in the following triangle:

In this triangle, the opposite side of 30° angle is half of the hypotenuse. And the opposite side of 60° is opposite of $30° \times \sqrt{3}$

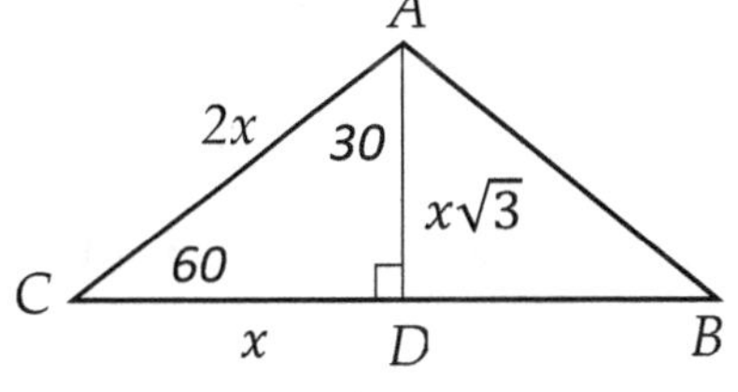

CD = 8, then AD = $8 \times \sqrt{3}$

Area of the triangle ABC is:

$\frac{1}{2}\ AD \times BC = \frac{1}{2}\ 8\sqrt{3} \times 16 = 64\sqrt{3}$

23) Answer: C.

Substituting 6 for y in $y = cx^2 + d$ gives $6 = cx^2 + d$ which can be rewritten as $6 - d = cx^2$. Since $y = 6$ is one of the equations in the given system, any solution x of $6 - d = cx^2$ corresponds to the solution $(x, 6)$ of the given system. Since the square of a real number is always nonnegative, and a positive number has two square roots, the equation $6 - d = cx^2$ will have two solutions for x if and only if (1) $c > 0$ and $d <$

6 or (2) $c < 0$ and $d > 6$. Of the values for c and d given in the choices, only $c = -1$, $d = 7$ satisfy one of these pairs of conditions. Alternatively, if $c = -1$ and $d = 7$, then the second equation would be $y = -x^2 + 7$; The equation has two real answer.

24) Answer: D.

$3x$ and z are colinear. $2y$ and $5x$ are colinear. Therefore,

$3x + z = 2y + 5x$, subtract x from both sides, then, $z = 2y + 2x$

25) Answer: D.

Here we can substitute 2 for x in the equation.

Thus, $y - 6 = 2(2 + 9)$, $y - 6 = 22$

Adding 2 to both side of the equation: $y = 22 + 6$, $y = 28$

26) Answer: C.

It is given that $g(6) = 12$. Therefore, to find the value of $f(g(6))$, substitute for $g(6)$;

$f(g(6)) = f(12) = 27$.

27) Answer: C.

The equation of a circle with center (h, k) and radius r is$(x - h)^2 + (y - k)^2 = r^2$. To put the equation $x^2 + y^2 + 2x + 6y = 2$ in this form, complete the square as follows:

$x^2 + y^2 + 2x + 6y = 2$

$(x^2 + 2x) + (y^2 + 6y) = 2$

$(x^2 + 2x + 1) - 1 + (y^2 + 6y + 9) - 9 = 2$

$(x + 1)^2 + (y + 3)^2 = 12$

$(x + 1)^2 + (y + 3)^2 = (\sqrt{12})^2$

Therefore, the radius of the circle is $\sqrt{12}$.

28) Answer: B.

The equation can be rewritten as $3a - c = ba$ →(divide both sides by a)$3 - \frac{c}{a} = b$, since $c > 0$ and $a < 0$, the value of $-\frac{c}{a}$ is positive. Therefore, 2 plus a positive number is positive. b must be greater than 3; $b > 3$.

29) Answer: C.

Let's review the options:

I. $|a| < 3 \rightarrow -3 < a < 3$

Multiply all sides by b. Since $b > 0 \rightarrow -3b < ba < 3b$

II. Since, $-3 < a < 3, and\ a < 0 \rightarrow -a > a^2 > a$ (plug in $-\frac{1}{3}$, check!)

III. $-6 < 2a < 6, multiply\ all\ sides\ by\ 2, then$:

$-6 < 2a < 6$, subtract 2 from all sides, then:

$-6 - 2 < 2a - 2 < 6 - 2 \rightarrow -8 < 2a - 2 < 4$

I and III are correct.

30) Answer: B.

By definition, the sine of any acute angle is equal to the cosine of its complement. Since, angle A and B are complementary angles, therefore:

$sin\ A\ =\ cos\ B$

31) Answer: 8.

Squaring both sides of the equation gives $4m + 32 = m^2$

Subtracting both sides by $4m + 32$ gives us the equation $m^2 - 4m - 32 = 0$

Here you can solve the quadratic by factoring to get $(m - 8)(m + 4) = 0$

For the expression $(m - 8)(m + 4)$ to equal zero, $m = 8$ or $m = -4$

Since m is a positive integer, 8 is the answer.

32) Answer: 6.

Since we are dealing with an absolute value, $f(a) = 20$means that either $16 - a^2 = 20$ or $16 - a^2 = -20$

Let's start with the positive value (58) and see what we get. If $16 - a^2 = 20$, then $a^2 = -4 \Rightarrow a = \sqrt{-4}$

On the other hand, if $16 - a^2 = -20$, then $a^2 = 36$

Taking the square root, we get $a = 6\ or - 6$

Notice that the question states that a is a positive integer, therefore the answer is 6.

33) Answer: 56.52.

The equation of a circle with center (h, k) and radius r is$(x-h)^2+(y-k)^2=r^2$. To put the equation $x^2+y^2+6x-8y+7=0$ in this form, complete the square as follows:

$x^2+y^2+6x-8y+7=0 \Rightarrow (x^2+6x)+(y^2-8y)+7=0$

$(x^2+6x+9)-9+(y^2-8y+16)-16+7=0$

$(x+3)^2+(y-4)^2-16+7-9=0 \Rightarrow (x+3)^2+(y-4)^2-18=0$

$(x+3)^2+(y-4)^2=18 \Rightarrow r^2$ equals 18. Then, $r=\sqrt{18}$

The radius of the circle is $\sqrt{18}$. Thus, the area of the circle is:

$A=\pi r^2=3.14(\sqrt{18})^2=3.14\times 18=56.52$

Round the answer to one decimal place to get 56.52.

34) Answer: 500.

One of the four numbers is x; let the other three numbers be y, z and w. Since the sum of four numbers is 900, the equation $x+y+z+w=900$ is true. The statement that x is 25% more than the sum of the other three numbers can be represented as

$x=1.25(y+z+w)$ or $\frac{x}{1.25}=y+z+w \rightarrow \frac{4x}{5}=y+z+w$

Substituting the value $y+z+w$ in the equation $x+y+z+w=900$

gives $x+\frac{4x}{5}=900 \rightarrow \frac{9x}{5}=900 \rightarrow 9x=4{,}500 \rightarrow x=\frac{4{,}500}{9}=500$

35) Answer: 7.

Let x represent the number of liters of the 40% solution.

The amount of salt in the 40% solution $(0.40x)$ plus the amount of salt in the 60% solution (0.60) × (3) must be equal to the amount of salt in the 25% mixture (0.25 × (x + 3)). Write the equation and solve for x.

$0.40x+0.60(3)=0.25(x+3) \rightarrow 0.40x+1.8=0.25x+0.75$

$\rightarrow 0.25x-0.40x=1.8-0.75 \rightarrow 0.15x=1.05 \rightarrow x=\frac{1.05}{0.15}=7$

36) Answer: 436.

This is a simple matter of substituting values for variables.

We are given that the 30 cars were washed today, therefore we can substitute that for a. Giving us the expression $\frac{30(30)-480}{30}+b$

We are also given that the profit was \$450, which we can substitute for $f(a)$. Which gives us the equation $450=\frac{30(30)-480}{30}+b$

Simplifying the fraction gives us the equation $450=14+b$

And subtracting both sides of the equation by 14 gives us $b=436$, which is the answer.

37) Answer: -7.

The function $f(x)$ is undefined when the denominator of $\frac{1}{(x+4)^2+6(x+4)+9}$ is equal to zero. The expression $(x+4)^2+3(x+4)+9$ is a perfect square.

$(x+4)^2+6(x+4)+9=((x+4)+3)^2$ which can be rewritten as $(x+7)^2$. The expression $(x+7)^2$ is equal to zero if and only if $x=-7$. Therefore, the value of x for which $f(x)$ is undefined is -7.

38) Answer: 50.

The area of ΔBED is 45, then: $\frac{10\times AB}{2}=45\rightarrow 5\times AB=45\rightarrow AB=9$

The area of ΔBDF is 40, then: $\frac{5\times BC}{2}=40\rightarrow 5\times BC=80\rightarrow BC=16$

The perimeter of the rectangle is $=2\times(9+16)=50$

SAT Mathematics

Practice Tests 2:

Section 1

1) Answer: C.

To rewrite $\frac{-2+5i}{1+2i}$ in the standard form $a + bi$, multiply the numerator and denominator of $\frac{-2+5i}{1+2i}$ by the conjugate, $1 - 2i$. This gives $\left(\frac{-2+5i}{1+2i}\right)\left(\frac{1-2i}{1-2i}\right) = \frac{-2+4i+5i-10i^2}{1^2-(2i)^2}$.

Since $i^2 = -1$, this last fraction can be rewritten as $\frac{-2+9i-10(-1)}{1-4(-1)} = \frac{8+9i}{5}$.

2) Answer: C.

First find the value of b, and then find $f(4)$. Since $f(5) = 75$, substuting 5 for x and 75 for $f(x)$ gives $75 = b(5)^2 + 5(5) = 25b + 25$ Solving this equation gives $b = 2$.

Thus $f(x) = 2x^2 + 5x$,

$f(4) = 2(4)^2 + 5(4) \rightarrow f(4) = 32 + 20; f(4) = 52$

3) Answer: D.

$\begin{cases} 2x + 4y = -8 \\ 8x + 15y = -30 \end{cases}$ ⇒ Multiplication (–4) in first equation ⇒ $\begin{cases} -8x - 16y = 32 \\ 8x + 15y = -30 \end{cases}$

Add two equations together ⇒ $-y = 32 - 30 \Rightarrow y = -2$ then: $x = 0$

4) Answer: D.

Identify the input value. Since the function is in the form $f(x)$ and the question asks to calculate $f(5)$, the input value is two.

$f(5) \rightarrow x = 5$, Using the function, input the desired x value.

Now substitute 5 for every x in the function. $f(x) = x^2 - 4x + 8$,

$f(5) = (5)^2 - 4(5) + 8, \quad f(5) = 25 - 20 + 8; f(5) = 13$

5) Answer: D.

Since $N = 6$, substitute 6 for N in the equation $\frac{x-7}{5} = N$, which gives $\frac{x-7}{5} = 6$.

Multiplying both sides of $\frac{x-7}{5} = 6$ by 5 gives $x - 7 = 30$ and then adding 7 to both sides of $x - 7 = 30$ then, $x = 37$.

6) Answer: C.

$b^{\frac{m}{n}} = \sqrt[n]{b^m}$ For any positive integers m and n. Thus, $b^{\frac{5}{7}} = \sqrt[7]{b^5}$.

7) Answer: B.

The total number of pages read by Sara is 6 (hours she spent reading) multiplied by her rate of reading: $N\,\frac{pages}{hour} \times 6 \text{ hours} = 6N$

Similarly, the total number of pages read by Mary is 11 (hours she spent reading) multiplied by her rate of reading: $M\,\frac{pages}{hour} \times 11 \text{ hours} = 11M$ the total number of pages read by Sara and Mary is the sum of the total number of pages read by Sara and the total number of pages read by Mary: $6N + 11M$.

8) Answer: C.

The problem asks for the sum of the roots of the quadratic equation $2n^2 + 26n + 80 = 0$. Dividing each side of the equation by 2 gives $n^2 + 13n + 40 = 0$. If the roots of $n^2 + 13n + 40 = 0$ are n_1 and n_2, then the equation can be factored as $n^2 + 13n + 40 = (n - n_1)(n - n_2) = 0$.

Looking at the coefficient of n on each side of $n^2 + 13n + 40 = (n + 8)(n + 5)$ gives $n = -8$ or $n = -5$, then, $-8 + (-5) = -13$

9) Answer: D.

The x-intercepts of the parabola represented by $y = x^2 - 8x + 15$ in the xy-plane are the values of x for which y is equal to 0.

The factored form of the equation, $y = (x - 3)(x - 5)$, shows that y equals 0 if and only if $x = 3$ or $x = 5$. Thus, the factored form $y = (x - 3)(x - 5)$, displays the x-intercepts of the parabola as the constants 3 and 5.

10)Answer: A.

To solve this problem first solve the equation for c. $\frac{c}{b} = 7$

Multiply by b on both sides. Then: $b \times \frac{c}{b} = 7 \times b \rightarrow c = 7b$

Now to calculate $\frac{21b}{c}$, substitute the value for c into the denominator and simplify:

$\frac{21b}{c} = \frac{21b}{7b} = \frac{21}{7} = 3$

11)Answer: B.

Simplify the numerator. $\frac{3x+(4x)^2+(5x)^3}{x} = \frac{3x+4^2x^2+5^3x^3}{x} = \frac{3x+16x^2+125x^3}{x}$

Pull an x out of each term in the numerator; $\frac{x(3+16x+125x^2)}{x}$

The x in the numerator and the x in the denominator cancel:

$3 + 16x + 125x^2 = 125x^2 + 16x + 3$

12)Answer: B.

The equation $\frac{a-b}{b} = \frac{5}{12}$ can be rewritten as $\frac{a}{b} - \frac{b}{b} = \frac{5}{12}$, from which it follows that $\frac{a}{b} - 1 = \frac{5}{12}$, or $\frac{a}{b} = \frac{5}{12} + 1 = \frac{17}{12}$.

13)Answer: A.

First write the equation in slope intercept form.

Add $4x$ to both sides to get $16y = 4x + 48$.

Now divide both sides by 16 to get $y = \frac{1}{4}x + 3$.

The slope of this line is $\frac{1}{4}$, so any line that also has a slope of $\frac{1}{4}$ would be parallel to it.

Only choice A has a slope of $\frac{1}{4}$.

14)Answer: C.

If $ax - b$ is a factor of $g(x)$, then $g\left(\frac{b}{a}\right)$ must equal 0. Based on the table $g(0) = -6$.

Therefore, $a(0) - b = -6 \Rightarrow b = 6$ then $g(6) = 0 \Rightarrow 6a - 6 = 0 \Rightarrow a = 1$.

$g(x) = x - 6$ must be a factor of $g(x)$.

15)Answer: B.

$4\blacksquare 5 = \sqrt{4^2 + 4(5)} = \sqrt{16 + 20} = \sqrt{36} = 6$

16)Answer: 2.5 or $2\frac{1}{2}$.

One-degree equals $\frac{\pi}{180}$.

The angle α in radians is equal to the angle α in degree times π constant divided by 180 degrees. Then: $1\ degree = \frac{\pi}{180} \rightarrow 450\ degrees = \frac{450\pi}{180} = 2.5\pi$,

$2.5\pi = x\pi \rightarrow x = 2.5$

17) Answer: 27.5 m.

The rate of construction company= $\frac{40 \text{ cm}}{1 \text{ min}} = 40 \, \frac{cm}{min}$

Height of the wall after 55 minutes $= \frac{40 \text{ cm}}{1 \text{ min}} \times 55 \text{ min} = 2{,}200 \text{ cm}$

Let x be the height of wall, then $\frac{4}{5}x = 2{,}200 \text{ cm} \rightarrow x = \frac{5 \times 2{,}200}{4}$

$\rightarrow x = 2{,}750 \text{ cm} = 27.5\, m$

18) Answer: $\frac{1}{25}$ or 0.04

Write the ratio of $5a$ to $9b$; $\frac{5a}{9b} = \frac{1}{45}$

Use cross multiplication and then simplify. $5a \times 45 = 9b \times 1 \rightarrow 225a = 9b$

$\rightarrow a = \frac{9b}{225} = \frac{b}{25}$

Now, find the ratio of a to b. $\frac{a}{b} = \frac{\frac{b}{25}}{b} \rightarrow \frac{b}{25} \div b = \frac{b}{25} \times \frac{1}{b} = \frac{b}{25b} = \frac{1}{25}$

19) Answer: 162.

The sum of all angles in a quadrilateral is 360 degrees. Let x be the smallest angle in the quadrilateral. Then the angles are: $2x, 3x, 6x, 9x$

$2x + 3x + 6x + 9x = 360 \rightarrow 20x = 360 \rightarrow x = 18$

The angles in the quadrilateral are 36°, 54°, 108°, and 162°

20) Answer: $\frac{4}{5}$

$\tan A = \frac{opposite}{adjacent}$, and $\tan A = \frac{9}{12}$, therefore, the opposite side of the angle A is 12 and the adjacent side is 9. Let's draw the triangle.

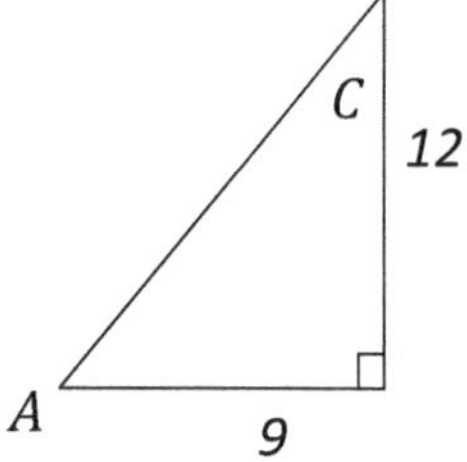

Using Pythagorean theorem, we can solve for the hypotenuse:

$a^2 + b^2 = c^2 \rightarrow 12^2 + 9^2 = c^2 \rightarrow 144 + 81 = c^2 \rightarrow c = 15$

$\cos C = \frac{adjacent}{hypotenuse} = \frac{12}{15} = \frac{4}{5}$

SAT Mathematics

Practice Tests 2:

Section 2

1) Answer: C.

If $\beta = 5\gamma$, then multiplying both sides by 3 gives $3\beta = 15\gamma$.

$\alpha = 3\beta$, thus $\alpha = 15\gamma$. Multiply both sides of the equation by 3 gives $3\alpha = 45\gamma$.

2) Answer: B.

From the choices provided, plugin the values of a and b into both inequalities and check.

A. (7, 3); and $a + b = 7 + 3 = 10 < 12$

B. (8, 5); and $a + b = 8 + 5 = 13 < 12$

C. (4, 0); and $a + b = 4 + 0 = 4 < 12$

D. $(6, 2)$; and $a - b = 6 - 2 = 4 > 2$; and $a + b = 6 + 2 = 8 < 10$

For choice B, 13 is not less than 12. Therefore, choice B does not provide the correct values of a and b.

3) Answer: A.

$\begin{cases} x + 3y = 14 \\ 5x + 8y = 49 \end{cases} \rightarrow$ Multiply the top equation by -5 then,

$\begin{cases} -5x - 15y = -70 \\ 5x + 8y = 49 \end{cases} \rightarrow$ Add two equations.

$-7y = -21 \rightarrow y = 3$, plug in the value of y into the first equation: $x + 3y = 14 \rightarrow$ $x + 3(3) = 14 \rightarrow x + 9 = 14$

Add -9 from both sides of the equation.

Then: $x + 9 = 14 \rightarrow x = 5$

4) Answer: A.

There are 17 floors, x rooms in each floor, and y chairs per room. If you multiply 17 floors by x, there are $17x$ rooms in the hotel. To get the number of chairs in the hotel, multiply $17x$ by y. $17xy$ is the number of chairs in the hotel.

5) Answer: B.

When 7 times the number x is added to 4, the result is $4 + 7x$. Since this result is equal to 53, the equation $4 + 7x = 53$ is true. Subtracting 4 from each side of $4 + 7x = 53$ gives $7x = 49$, and then dividing both sides by 7 gives $x = 7$. Therefore, 3 times x added to 8, or $8 + 3x$, is equal to $8 + 3(7) = 29$.

6) Answer: B.

Fining g in term of h, simply means "solve the equation for g".

To solve for g, isolate it on one side of the equation.

Since g is on the left-hand side, just keep it there.

Subtract both sides by $9h$. $9h + g - 9h = 11h + 5 - 9h$

And simplifying makes the equation $g = 2h + 5$, which happens to be the answer.

7) Answer: D.

The description $4 + 5x$ *is* 12 more than 27 can be written as the equation $4 + 5x = 12 + 27$, which is equivalent to $4 + 5x = 39$. Subtracting 4 from each side of $4 + 5x = 39$ gives $5x = 35$.

Since $15x$ is 3 times $5x$, multiplying both sides of $5x = 35$ by 3 gives $15x = 105$

8) Answer: C.

Plugin the values of x in the choices provided. The points are $(2, -9), (3, -14), and\ (4, -19)$

For $(2, -9)$ check the options provided:

A. $g(x) = x + 5 \rightarrow -9 = (2) + 5 \rightarrow -9 = 6$ This is NOT true.

B. $g(x) = x - 11 \rightarrow -9 = (2) - 11 = -9$

Check $(3, -14)$: $-14 = 3 - 11 \rightarrow -14 = -8$ This is NOT true.

C. $g(x) = -5x + 1 \rightarrow -9 = -5(2) + 1 \rightarrow -9 = -9$ This is true.

Check $(3, -14)$: $-14 = -5(3) + 1 = -14$ This is true.

Check $(4, -19)$: $-19 = -5(4) + 1 = -19$ This is true.

D. $g(x) = -5x - 1 \rightarrow -9 = -5(2) - 1 \rightarrow -9 = -11$ This is NOT true.

From the choices provided, only choice C is correct.

9) Answer: C.

Since all the numbers in the set are either integers or non-integers and the ratio of integers to non-integers is 9:11, one out of two numbers are integer. Therefore, the ratio of integers to all number is 9:11 that is equal $\frac{1}{20} \times 100 = 5\%$.

10) Answer: B.

To find the y −intercept of a line from its equation, put the equation in slope-intercept form: $x - 3y = 18 \Longrightarrow -3y = -x + 18 \Rightarrow 3y = x - 18 \Rightarrow y = \frac{1}{3}x - 6$

The y −intercept is what comes after the x. Thus, the y −intercept of the line is -6.

11) Answer: C.

The sum of the two polynomials is $(8x^2 + 5x - 11) + (3x^2 - 4x + 15)$

This can be rewritten by combining like terms:

$(8x^2 + 5x - 11) + (3x^2 - 4x + 15) = (8x^2 + 3x^2) + (5x - 4x) + (-11 + 15) =$
$11x^2 + x + 4$

12) Answer: D.

Percentage of men in city C = $\frac{810}{1,585} \times 100 = 51.10\%$

Percentage of men in city B = $\frac{460}{870} \times 100 = 52.87\%$

Percentage of men in city C to percentage of men in city B: $\frac{51.10}{52.87} = 96.65$

13) Answer: C.

Ratio of women to men in city A: $\frac{620}{650} = 0.95$

Ratio of women to men in city B: $\frac{410}{460} = 0.89$

Ratio of women to men in city C: $\frac{775}{810} = 0.96$

Ratio of women to men in city D: $\frac{648}{670} = 0.97$

14) Answer: B.

Let the number of women should be added to city D be x, then:

$$\frac{648 + x}{670} = 1.7 \rightarrow 648 + x = 670 \times 1.7 = 1,139 \rightarrow x = 491$$

15) Answer: C.

Adding both side of $8a - 7 = 33$ by 7 gives $8a = 40$

Divide both side of $8a = 40$ by 8 gives $a = 5$, then $5a = 5(5) = 25$

16) Answer: B.

The easiest way to solve this one is to plug the answers into the equation.

When you do this, you will see the only time $x = x^{-3}$ is when $x = 1$ or $x = 0$; Only $x = 1$ is provided in the choices.

17) Answer: A.

First find a common denominator for both of the fractions in the expression $\frac{8}{x^4} + \frac{3x-4}{x^5}$.

of x^5, we can combine like terms into a single numerator over the denominator:

$\frac{8}{x^4} + \frac{3x-4}{x^5} = \frac{8x+3x-4}{x^5} = \frac{11x-4}{x^5}$

18) Answer: D.

To find the average of three numbers even if they're algebraic expressions, add them up and divide by 3. Thus, the average equals:

$$\frac{(-3x+6)+(-5x-8)+(5x+14)}{3} = \frac{-3x+12}{3} = -x+4$$

19) Answer: C.

Adding the two equations side by side eliminates y and yields $x = 8$.

$\frac{5}{2}y = 6,\ x - \frac{5}{2}y = 2, \quad \rightarrow x - 6 = 2 \rightarrow x = 8$

20) Answer: D.

Let's find the vertex of each choice provided:

A. $y = 4x^2 - 4$ The vertex is: $(0, -4)$

B. $y = -4x^2 + 4$ The vertex is: $(0, 4)$

C. $y = x^2 + 6x - 8$

The value of x of the vertex in the equation of a quadratic in standard form is:

$x = \frac{-b}{2a} = \frac{-6}{2}$

(The standard equation of a quadratic is: $ax^2 + bx + c = 0$).

The value of x in the vertex is 4 not -3.

D. $y = 5(x-4)^2 - 4$

Vertex form of a parabola equation is in form of $y = a(x-h)^2 + k$, where (h, k) is the vertex. Then $h = 4$ and $k = -4$. (This is the answer)

21) Answer: C.

Let l and w be the length and width, respectively, of the original rectangle. The area of the original rectangle is $A = lw$. The rectangle is altered by increasing its length by 22 percent and decreasing its width by s percent; thus, the length of the altered rectangle is $1.22l$, and the width of the altered rectangle is $\left(1 - \frac{s}{100}\right)w$.

The alterations decrease the area by 8.5 percent, so the area of the altered rectangle is $(1 - 0.085)\,A = 0.915A$. The altered rectangle is the product of its length and width, therefore $0.915A = (1.22l)\left(1 - \frac{s}{100}\right)w$

Since $A = lw$, this equation can be rewritten as $0.915A = (1.22)\left(1 - \frac{s}{100}\right)lw = (1.22)\left(1 - \frac{s}{100}\right)$, from which it follows that $0.915 = (1.22)\left(1 - \frac{s}{100}\right)$, divide both sides of the equation by 1.22. Then: $0.75 = 1 - \frac{s}{100}$

Therefore, $\frac{s}{100} = 0.25$ and therefore the value of s is 25%.

22) Answer: B.

According to the properties of exponents, you can rewrite the equation as $\frac{1}{\sqrt[2]{b^7}} = x$

(Remember that $a^{-n} = \frac{1}{a^n}$ and $a^{\frac{1}{n}} = \sqrt[n]{a}$)

Proceed to solve for b first by squaring both sides, which gives $\frac{1}{b^7} = x^2$.

And then multiplying both sides by b^7 to find $1 = b^7x^2$. finally, dividing both sides by x^2 isolates the desired variable.

$\frac{1}{x^2} = b^7 \rightarrow \sqrt[7]{\frac{1}{x^2}} = b$

23) Answer: B.

Because each child ticket costs \$2 and each adult ticket costs \$5, the total amount, in dollars, that John spends on x children tickets and 4 adult ticket is $2(x) + 4(5)$. Because

John spends at least \$23 but no more than \$27 on the tickets, you can write the compound inequality $2x + 20 \geq 23$ and $2x + 20 \leq 27$.

Subtracting 20 from each side of both inequalities and then dividing each side of both inequalities by 2 gives $x \geq 1.5$ and $x \leq 3.5$. Thus, the value of x must be an integer that is both greater than or equal to 1.5 and less than or equal to 3.5. Therefore, $x = 2$ or $x = 3$. Either 2 or 3 may be gridded as the correct answer.

24) Answer: A.

Since a box of pen costs \$3, then $3p$ Represents the cost of p boxes of pen.

Multiplying this number times 1.038 will increase the cost by the 3.8% for tax.

Then add the \$8 shipping fee for the total: $1.038(3p) + 8$

25) Answer: D.

Rate of change (growth or x) is 9 per week. $63 \div 9 = 7$

Since the plant grows at a linear rate, then the relationship between the height (y) of the plant and number of weeks of growth (x) can be written as: $y(x) = 7x$

26) Answer: C.

If the central angle measures 120 degrees, divide the 360 total degrees in the circle by 120: $\frac{360}{120} = 3$

Multiply this by the measure of the corresponding arc to find the circumference of the circle: $6 \times 3 = 18$

Use the circumference formula to find the radius, then use the radius to find the area.

$$\text{C} = 2\pi r \to 18 = 2\pi r \to \frac{18}{2\pi} = r \to \frac{9}{\pi} = r$$

$$\text{A} = \pi r^2 \to A = \pi \times \left(\frac{9}{\pi}\right)^2 \to A = \frac{81}{\pi}$$

27) Answer: C.

Add the first 4 numbers. $52 + 44 + 32 + 59 = 187$

To find the distance traveled in the next 6 hours, multiply the average by number of hours. Distance = Average × Rate = $51 \times 6 = 306$

Add both numbers. $187 + 306 = 493$

28) Answer: C.

Since function $g(x)$ is a linear function, we can write it in the form of $g(x) = mx + d$

We start by finding the slope, m. The slope is given by $\frac{d_1 - d_2}{a_1 - a_2}$

Choosing two values $(1, -9)$ and $(-4, 16)$ gives us $m = \frac{16-(-9)}{-4-1} = -5$

Next, we solve for d in the equation $g(x) = -5x + d$.

Looking at the first set of numbers on the table, we can see that when $x = 1, g(x) = -9$; Substituting these values in the equation $g(x) = -5x + d$ gives us $-9 = (-5)(1) + d$

And simplifying gives us $d = -4$. Therefore, the linear equation of the function is: $g(x) = -5x - 4$; But that's not one of the choices!

The table shows that $g(x) = b, x = a$ then we can merely substitute them, which gives us the equation $b = -5a - 4$

29) Answer: A.

It is given that the function $g(x)$ passes through the point (2, 4). Thus, if $x = 2$, the value of $g(x)$ is 4 (since the graph of g in the xy-plane is the set of all points $(x, g(x))$.

Substituting 2 for x and 4 for $g(x)$ in $g(x) = 2x^2 + cx - 8$ gives $4 = 2(2)^2 + c(2) - 8$

Solve this equation for c. Then: $4 = 8 + 2c - 8$

$4 - 8 + 8 = 2c \rightarrow c = 2$

30) Answer: B.

If the value of $|x - 9| + 9$ is equal to 0, then $|x - 9| + 9 = 0$. Subtracting 9 from both sides of this equation gives $|x - 9| = -9$. The expression $|x - 9|$ on the left side of the equation is the absolute value of $x - 9$, and the absolute value can never be a negative number.

Thus $|x - 9| = -9$ has no solution. Therefore, there are no values for x for which the value of $|x - 9| + 9$ is equal to 0.

31) Answer: 5 or 8.

First factor the function: $f(x) = 3x^3 - 39x^2 + 120x = 3x\,(x - 5)(x - 8)$

To find the zeros, $f(x)$ should be zero. $f(x) = 3x\,(x-5)(x-8) = 0$

Therefore, the zeros are: $x = 0, (x-5) = 0 \Rightarrow x = 5$,

$(x-8) = 0 \Rightarrow x = 8$

32) Answer: 380.

To determine the number of drinks sold, write and solve a system of two equations.

Let x be the number of salads sold and let y be the number of drinks sold. Then: $x + y = 400$

Since each salad cost \$10, each drink cost \$5, and the total revenue was \$2,100, the equation $10x + 5y = 2{,}100$ is true.

The equation $x + y = 400$ is equivalent to $5x + 5y = 2{,}000$, and subtracting each side of $5x + 5y = 2{,}000$ from the respective side of $10x + 5y = 2{,}100$ gives $5x = 100$. Therefore, the number of salads sold, x was $x = 20$ and the number of drinks sold was $y = 400 - 20 = 380$.

33) Answer: 64.

$\text{average} = \frac{\text{sum of terms}}{\text{number of terms}} \Rightarrow 65 = \frac{\text{sum of terms}}{40} \Rightarrow \text{sum} = 65 \times 40 = 2{,}600$

The difference of 75 and 35 is 40. Therefore, 40 should be subtracted from the sum.

$2{,}600 - 40 = 2{,}560$

$\text{mean} = \frac{\text{sum of terms}}{\text{number of terms}} \Rightarrow \text{mean} = \frac{2{,}560}{40} = 64$

34) Answer: $\frac{16}{23}$.

Since the outcomes are mutually exclusive. Then, the sum of probabilities of all outcomes equals to 1. Therefore: $n + \frac{n}{4} + \frac{n}{8} + \frac{n}{16} = 1$

Find a common denominator and solve for n.

$n + \frac{n}{4} + \frac{n}{8} + \frac{n}{16} = 1 \rightarrow \frac{16n}{16} + \frac{4n}{16} + \frac{2n}{16} + \frac{1n}{16} = 1 \rightarrow \frac{23n}{16} = 1 \rightarrow 23n = 16 \rightarrow n = \frac{16}{23}$

35) Answer: 184.

The first term in the sequence is 4 and each term in this sequence is found by adding 6 to the preceding term. This is an arithmetic sequence. Given an arithmetic sequence with the first term a_1 and the common difference d, the n^{th}(or general) term is given by:$a_n = a_1 + (n-1)d$, Then: $a_{31} = 4 + (31-1)6 = 184$

36) Answer: 65.

The speed of car A is 50 mph, and the speed of car B is 80 mph. When both cars drive in a straight line toward each other, the distance between the cars decreases at the rate of 130 miles per hour: 50 + 80 = 130

30 minutes is half of an hour. Therefore, they will be 65 miles apart 30 minutes before they meet. $\frac{1}{2} \times 130 = 65$

37) Answer: 11.

To solve for the inverse function, first replace $f(x)$ with y. Then, solve the equation for x and after that replace every x with a y and replace every y with an x. Finally, replace y with $f^{-1}(x)$. $f(x) = \frac{2x-7}{3} \Rightarrow \text{y} = \frac{2x-7}{3} \Rightarrow 3\text{y} = 2x - 7 \Rightarrow 3\text{y} + 7 = 2x \Rightarrow \frac{3\text{y}+7}{2} = x; f^{-1}(x) = \frac{3\text{x}+7}{2} \Rightarrow f^{-1}(5) = \frac{3(5)+7}{2} = \frac{22}{2} = 11$

38) Answer: 14.

Since the angles are acute and $sin(a°) = cos(b°)$, the angles are complementary angles and $a + b = 90$. Substituting $2n + 3$ for a and $5n - 11$ for b gives, $(2n + 3) + (5n - 11) = 90$ which simplifies to $7n - 8 = 90$.

Therefore, $7n = 98$, and $n = 14$

"End"

www.ingramcontent.com/pod-product-compliance
Lightning Source LLC
LaVergne TN
LVHW061240100826
845148LV00008B/994

* 9 7 8 1 6 3 6 2 0 0 4 6 0 *